Pongsak Chaisuparasmikul

Simplified Building Energy Modeling Tool for Architects

Simplified Software Tool for Architects and Energy Efficient Building Design Professionals

VDM Verlag Dr. Müller

Imprint

Bibliographic information by the German National Library: The German National Library lists this publication at the German National Bibliography; detailed bibliographic information is available on the Internet at http://dnb.d-nb.de.
    Any brand names and product names mentioned in this book are subject to trademark, brand or patent protection and are trademarks or registered trademarks of their respective holders. The use of brand names, product names, common names, trade names, product descriptions etc. even without a particular marking in this works is in no way to be construed to mean that such names may be regarded as unrestricted in respect of trademark and brand protection legislation and could thus be used by anyone.

Cover image: www.purestockx.com

Publisher:
VDM Verlag Dr. Müller Aktiengesellschaft & Co. KG, Dudweiler Landstr. 125 a, 66123 Saarbrücken, Germany,
Phone +49 681 9100-698, Fax +49 681 9100-988,
Email: info@vdm-verlag.de

Copyright © 2008 VDM Verlag Dr. Müller Aktiengesellschaft & Co. KG and licensors
All rights reserved. Saarbrücken 2008

Produced in USA and UK by:
Lightning Source Inc., La Vergne, Tennessee, USA
Lightning Source UK Ltd., Milton Keynes, UK
BookSurge LLC, 5341 Dorchester Road, Suite 16, North Charleston, SC 29418, USA

ISBN: 978-3-639-00977-4

Pongsak Chaisuparasmikul

Simplified Building Energy Modeling Tool for Architects

**This book is dedicated to my Lord Buddha,
my mother Somporn, and my sister Somchit**

About the Author:

Pongsak has PhD Degree Architecture from Illinois Institute of Technology. He has practiced as an architect and planner since 1982, for 16 years, and taught at the two accredited architectural school in Thailand for 6 years. In 1987 he has received architectural licensure to practice as a professional architect in Thailand. He has designed and consulted for a variety of challenging projects, range from multi-family residential and condominium, hotel, healthcare facilities, resorts and hospitality, office buildings, industrial, retail, and civic centers. He has established a reputation in his field and acknowledged in many ways through projects, competitions, publications, grants, invited presentations, teaching and guest lecturer. He has received citation of his work from Architecture Weeks (2006) admiration of his work from Director of Center for Sustainable Cities and Professor at School of Architecture, College of Design, University of Kentucky (2006), admiration of his research from Dean of the College o Architecture, Illinois Institute of Technology (2006); and admiration of his work's performance from Deputy Director, the Energy Research Institute (ERI) in Thailand (1998). Pongsak has done many high end sustainability, building energy modeling and simulation projects, and is listed as a DOE21.E consultant in the Simulation Research Group, Lawrence Berkeley National Laboratory. A mark of his specialization is his findings of the intimate relationship between building information model, sustainability theory and practice and energy simulation. One of his strong points is in his experience and ability to tie very different software programs together to share information and interact with one another in the meaningful way. He calls it "Interoperability of Model and Engine". He constructed the Sustainable City and Building Design Engine, a synthesis of modeling-simulation algorithms and databases. His research and software "Data Exchange Model and Software Interoperability to Improve the Energy-Efficient Building Design Process" has received 2007 research grant award from the Boston Society of Architects.

TABLE OF CONTENTS

viii

LIST OF TABLES

LIST OF FIGURES

Figure   Page

x

Figure   Page

Figure   Page

Figure   Page

xiv

PREFACE

Energy Modeler is an energy software program designed to study the relative change of energy uses (heating, cooling, and lighting loads) in different architectural design schemes. This research focuses on developing a tool to improve energy efficiency of the built environment. This energy-based interfacing program is intended to help architects, engineers, educators, students, building designers, major consumers of architectural services, and other professionals, whose work interfaces with that of architects, perceive, quickly visualize, and compare energy performance and savings of different design schemes.

Buildings are dynamic in nature and changeable over time, with many design variables involved. Architects really need energy-based rules or tools to assist them in the design process. Energy efficient design for sustainable solutions requires attention throughout the design process and is related to architectural solutions. Early involvement is the only guaranteed way of properly considering fundamental building design issues related to building site, form and exposure.

The research presents the methodology and process, which leads to the discussion of the research findings. The innovative work is to make these tools applicable to the earliest stage of design, where more informed analysis of possible alternatives could yield the most benefit and the greatest cost savings both economic and environmental. This is where computer modeling and simulation can really lead to better and energy efficient buildings. Both apply to internal environment and human comfort, and environmental impact from surroundings.

Pongsak Chaisuparasmikul

Chicago, Illinois

CHAPTER 1

INTRODUCTION

Energy Modeler is an energy software program designed to study the relative change of energy uses (heating, cooling, and lighting loads) in different architectural design schemes.

This research focuses on developing a tool to improve energy efficiency of the built environment. The research studied the impact of different architectural design response for two distinct global climates: temperate and tropical climatic zones. This energy-based interfacing program is intended to help architects, engineers, educators, students, building designers, major consumers of architectural services, and other professionals whose work interfaces with that of architects, perceive, quickly visualize, and compare energy performance and savings of different design schemes.

This program shows how energy efficiency is associated with environmental sustainability under varying conditions including, climatic condition, orientation, window surface ratio, glass type, and shading devices. Users can generate building and built environment conceptual design, factoring in the outside climate conditions. The program is intended for use during the initial architectural design phases where the site and building planning issue demonstrate the ability of architects to synthesize programmatic and environmental requirements into a coherent and aesthetic concept for the design of buildings and their placement on the site.

Its ease of use and ability to analyze alternatives in a short period of time, make it a useful educational tool as well as design or analysis tool. The program is developed using DOE2.1E energy simulation engine as well as Basic and Visual Basic programming. Intended users, a population which has the ability to influence sustainable building design and environmental consciousness, will now have the ability to identify the environmental impact potential and problems of their design, find and evaluate alternatives or options, and make decisions.

Often when designing buildings, architects may lack energy-based information of how building loads and energy consumption respond to the design of architectural massing or forms, building facing orientation, window- to- wall surface area ratio, glazing types, building exterior shading devices, or natural day lighting. A simplified energy tool could assist them in the design process. To produce energy-efficient projects, it requires attention to energy factors throughout the design process in the development of architectural solutions. Early involvement is helpful in the proper consideration of fundamental building design issues related to building site, form and exposure.

A successful project also requires advocacy by the owner, developer and the leader of the design team or architect, along with the engineer if energy efficient design is to be incorporated. The building design is dynamic in nature and changeable over time, with many design parameters involved. In order to meet energy efficient targets, it is necessary to model and simulate using state of the art and highly sophisticated computation analysis software to find the alternatives. Meaningful results can be obtained to confirm the design approach made by architects.

1.1 Scope of Work Statement

This research seeks to provide an integrated approach to the design of climatic responsive buildings and built environment sustainability by creating a software tool to enable architects and designers to visually compare energy efficiency. The work seeks the advancement of the building energy performance approach through which architects and building designers can find alternative solutions and make decisions, during the conceptual design process. This applied research seeks to develop an innovative knowledge to help architects and designers in decision making by revealing energy potential and problems, identifying opportunities that the architects and building designers faced in their design process.

The intent of this research is to provide a simplified energy model and program for the building envelope and skin parameters. For each of these parameters, there is the range of design input options so that as different options are selected, the user can visualize the energy performance change, and modify their design schemes in subsequent iterations.

This research is intended to assist architects and designers in decision-making, by providing the energy estimate graphs, charts and images so the architects can gain an understanding of energy parameters that have real impact on the energy performance while optimizing energy use.

1.2 Problem Definition: Technological Challenge to Sustainability

The buildings in which we live or work have a great impact on our natural environment. People working as architects, engineers, educators, developers and students play a key role in creation of the built environment. They have high influence on the economic and social prosperity; also have high potential for protecting the natural systems of the planet, and providing a higher quality of life for its people. Our natural environment has such a great impact on the buildings in which we live or places in which we work. Energy savings and consumption reductions in our buildings probably are the best indications of solutions to help environmental sustainability; by reducing the depletion of the world's fossil fuel (oil, natural gas, coal etc.). To build more power plants can lead to more deterioration of the natural environment. Architects, engineers, owners and developers should have a basic understanding not just a vague notion of how building, energy and environment really work together.

Architects when they set about designing an environmentally responsive building for an owner or the public, often lack the energy-based information and design tools to tell them whether the building loads and energy consumption are very responsive to the modifications that they made. Buildings are dynamic in nature and changeable over time, with many design variables involved. Architects really need energy-based rules or tools to assist them in the design process. Energy efficient design for sustainable solutions requires attention throughout the design process and is very related to architectural solutions. Early involvement is the only guaranteed way of properly considering fundamental building design issues related to building site, form and exposure.

It is at this intuitive stage that the greatest potential exists for energy performance efficiencies and environmental economies. The right orientation and fenestration choice can reduce the air-conditioning requirement. Incorporating passive solar elements and natural ventilation can help tremendously increase energy efficiency. In the design of the building envelope, building loads and energy consumption change in response to the change in the material properties and construction elements. It is much more difficult and costly to retrofit these features at a later stage of a project's development.

1.3 Hypotheses

This research suggests a method of programming, modeling and simulation of a series of design parameters through a representative data range, which can help designers to determine the design energy usage resulting from their design scheme. The diagrams contained herein suggest the pursuance of a holistic approach while exploring the integrated design process, to achieve a higher level of energy efficiency in architectural design.

1.4 Innovative Design Approach

The state of the art and highly sophisticated software integrates key building parameters. This software helps architects and designers solve the energy complex problems and find the environmental sound alternatives, without having to know all the theoretical concepts behind each knowledge area.

The use of energy efficient design tools and understanding building parametric performance (climatic condition, orientation, window ratio to fenestration, glass type, and shading devices) allows the building designers, educators and students to visualize the sustainable design alternatives that will consider thermal, visual and air quality aspects; foster frugal use of energy and other resources; and be gentle overall to the environment. Awareness of energy related design strategies could sustain the natural resources and help reduce energy use. The building design approach is based on a function of energy transforming system that is based on modeling concept. If the role of design tool is to serve design process at the conceptual intuitive phase, then this research will present ideas of what the approach should be; accompanied by the case scenarios that demonstrates their implementation.

The nature of architecture is that every problem has more than one solution. This is also the same for the energy and environmental design. Energy uses can be the function of many different parameters either on the architecture side: massing and shape ratio, climate location, orientation, envelope surface ratio, percentage of glazing and glass types, day lighting that allowed to penetrate into the space, and building exterior shading devices such as overhangs or light-shelves or on the engineer side: mechanical system and equipments. Evaluation of alternative solutions has long been an important part of the architectural or building design. If the designers and building maker are aware of these parameters and how these are performing, they can employ energy efficient strategies and logically approach the strategies of energy efficient design.

1.5 Research Objectives

This research describes the use of a simplified energy modeler, method and process for a hypothetical module. The architect with no experience in using computerized energy analysis tool will find this program very simple and helpful in determining energy performance. Variables can be changed for different design schemes, results can be quickly obtained, and energy problems and opportunities can be identified. This research has the following objectives:

1. Minimize the energy usage of the building and built environment.
2. Have reduced impact on the environment, resulting in lower use of the fossil fuels and natural resources.
3. Lead to the zero energy concept[1].
4. Be versatile and meet the future challenges.
5. Enable the study of different architectural design schemes for energy use impact. Enable better energy efficient and sustainable building design. Enable the identification of alternative ideas for potential, problems and opportunities with the energy design parameters.
8. To develop energy efficient design tool so that architects can find the alternative solutions and to making decision.
9. Ability to perform the whole building and facility and/or individual facility, quickly and many calculations.
10. To create the simplified program that can visualize the data in graphical or curve form[2] those are relevant to architectural schemes and can allow the understanding of energy changes much more quickly and accurately. Create an energy design-learning tool.

Study the significance and impact of input design parameters, and to identify important characteristics of the input and output variables. To determine the design sensitivity which can be sorted out of the important parameters or variables that has the impact validity on the energy usage or consumption of the building, to assess the significance and impact of input design parameters.

---

[1] Zero energy concepts mean no demand for energy from the plant or fossil fuel since the energy given from natural environment has taken care of all the requirements and comfort.
[2] Graphical or curve forms mean a line, plane or curved surface, which can be built into a performance curve and stored in the form of equations.

1.6 Research Methodology and Procedure

The research presents the methodology and process, which leads to the discussion of the research findings. The innovative work is to make these tools applicable to the earliest stage of design, where more informed analysis of possible alternatives could yield the most benefit and the greatest cost savings both economic and environmental. This is where computer modeling and simulation can really lead to better and energy efficient buildings. Both apply to internal environment and human comfort, and environmental impact from surroundings.

This research starts with a literature review of the historical development of energy and sustainability design. The next section is to survey the recent projects and software design and analysis tools that have incorporated or trying to bring the energy efficiency and sustainability measures into architectural design. Chapter 3 defines the module as the important concept of developing tools for architects and its energy parameters that are built inside that made the module flexible. Energy parameters are consisted of constant parameters and variable parameters. Chapter 4 describes the software tool called "Modeler" about its function, ability, and feature. Chapter 5 presents the case scenarios as the guidelines for how to use the software to solve the energy problems. Chapter 6 is conclusion of this research findings and contribution to the future development in this field.

The followings are the method and procedure undertaken:
1. Literature review of the contemporary architecture design, and their energy and environment related topics. Survey the Range of Possibilities that these envelope parameter option coefficients can covered
2. Define the energy parameter groups that are potential criteria that most architects have to deal with in the architectural design.
3. Identify potential energy conserving measures and strategies that fall into these scopes.
4. Create the Database for Simulation Model using ACCESS that help linking between the database and DOE2.1E input deck.

5. Design the data and Worksheet to define the database range.

6. Design and check the input parameters and make use of the database handling technique.

7. Write the BASIC program that identifies the parameters, coefficient of the energy model.

8. Develop the energy parameters in building simulation, implement using energy software engine DOE2.1E.

9. Data analysis was implemented Visual Basic (VB) to link the output results from DOE-2.1E and ACCESS.

10. Check the energy performance, load profile, base energy profile, and peak energy profile with the design.

11. Presenting the case studies in Chicago and Bangkok for applications and comparison with the example, problems, and suggested possible solutions as a mean to the method and process that can be used in any generic buildings.

Simulation is the process of developing a simplified model of a complex energy usage in a building, and using this model to analyze and predict the behavior of the energy performance. This research uses the energy modeling technique that understanding a building is a complex network of thermal resistance and capacitance linking different regions and representing conduction, convection, advection, radiation, and heat storage processes. The manner in which this network is treated mathematically, while simplifying boundary condition assumption might be made, and thus determining the flexibility of the modeling technique to emerge.

1.7 Assumptions and Limitations

This research is the methodology tool or manual guideline for strategic energy design efficiency and sustainability. It is intended to be used to test and visualize the relative energy performance of a number of design options, although can be regarded as a precise actual number because of the results from actual simulation. However, the precision of the actual number produced is less important than the energy model trends shown by the comparison of the parameters and variables. The research predicts the potential energy performance and design parameters of the module as representation of hypothetical buildings, assuming that the occupancy pattern and space system are function optimally.

The results should be interpreted in context with clear understanding of the implications. At the conceptual stage, building geometry and form is fluid and subject to constant change. The work is focused on the climate location of two cities, Chicago and Bangkok, representing two extreme climates. Although the weather data used are specific to these two cities, derived results can be representation of results for conceptual study of the locations within similar climate characteristics.

1.8 Research Findings, Outcomes and Results

In order to architect can make use of the data result. The visual representations of statistical data results are displayed as the graphs, charts, and tables, which is the graphical representation of the quantitative results in the program. By placing the statistical data in the appropriate context, architects can get useful input into their design schemes.

The ability to sort out these parameters or variables to how much of these fenestration or architectural options can be built into a problem formulation for identifying the optimal solution designs energy efficient building, and changing alternative of the variables.

In acting as a filter between internal and external climates, the building envelope will modify the environment in term of heat, light, sound, and in term of aesthetic aspect such as view and insolation. Some part of this environment can be regarded as desirable attributes, and some as undesirable. To make rational design decision, the architect must have some kind of information on the most of these environmental design consequences.

The outcomes are the resulting models and energy program that are developed to visualize the outcome diagrams, graphs, tables, and charts. The visual representation of statistical data and trend curves will give the primary energy consumption per design cost for each of the north, south, east, and west orientation of the façade. Curves and prediction equation will be presented for heating, cooling, lighting and total energy, which can be used as a simplified design tool for the designers to compare the design cost of the thermal and energy performance of different design schemes, giving the picture of the trend and relative importance of the energy components, and providing energy opportunities to the architectural design during the early design stage.

The energy program will help to see the prediction problem and energy opportunity that the designer can possibly attain. The guideline and recommendation can provide the architect the visual qualitative and quantitative information how the building reacts and various architectural design parameters interact when the design scheme is altered. The results can be used to evaluate the energy design performance in a number of options and to make comparison. This research also explains the precise relationship between the output results and the design input parameters analytically and explicitly, due to the complex effect of coupling by the building load, develop relationship between the simulation energy profile, actual energy profile, and base energy profile (weather non dependent loads).

CHAPTER 2

REVIEW AND SURVEY OF THE LITERATURE

"They consume one-third of their energy from fossil fuels. They use two-thirds of the electricity consumed in the United States and therefore account for two-thirds of carbon dioxide emissions. They consume one-sixth of the world's fresh water and one-quarter of the world's wood harvest. Their footprints often contaminate the landscape, creating unusable "brown fields." They are, of course, **buildings"**. (Hart, 1998)

"Environmental design replaces structure as the principal problem of architectural science" (Cowan H.J., 1966). In response, more than 20 years later, Manning writes: "Despite enormous amounts of research that has been undertaken into many aspects of building environment, and the store of knowledge that has accumulated, design of the environment too often appears to be a matter of chance. Users of today's new buildings are just as liable as were users of earlier buildings to be uncomfortable." (Manning P. 1987)

2.1 Importance of Energy and Sustainable Design to Architecture

Energy is a science of uncertainty and an art of probability. The nature of building technology design as applied research was put by Ove Arup, the founder of the Arub consulting engineer firm, when he said about architecture and engineering; "(Architecture and) Engineering are not a science. Science studies particular events to find general laws. (Architecture and) Engineering make use of this laws to solve particular problems. In this is most closely related to art or craft; as in art its problems are under-defined, there are many solutions, good, bad or indifferent. The art is, by a synthesis ends and means, to arrive as a good solution. This is a creativity, involving imagination, intuition and deliberately choice."

The building industry represents one of the largest enterprises in the country. For example, roughly one quarter of the assets of large U.S. corporations are tied up in buildings and land. About one third of all investment in the U.S. is for construction of commercial and residential buildings and more than one third of the total energy consumed in the U.S. is used in the building sector. Building design is full of uncertainty. Whether it is the architect uncertain of how much light a window might let in, or the mechanical engineer uncertain of exactly how much solar gain to allow for, this lack of surety results in a significant amount of over-design. It is standard practice throughout the building industry to include safety factors and margins for error within design calculations. Studies have shown that these are generally between 5 and 10%. (Parsloe, C.J., 1995, Brittain, J., 1997) However, research by the Building Services Research and Information Association (BSRIA) suggests that such small margins, when applied many times at different points within a design, can result in a cumulative design margin of up to 60%. (Race, G.L., 1997)

The building industry is one of the biggest economic sectors. In Europe alone, building related business represents a yearly turnover on the order of $460 billion. While buildings are the most energy intensive economic sector, primary energy consumption of buildings worldwide is close to 19 million barrels of oil per day and represents almost the entire daily production of OPEC countries. Buildings, at the same time, have a major effect on the environment, and the environment has a major effect on buildings.

Environmental quality of indoor space is a compromise between the properties of the building applied during its design and operations. As buildings have a long life of several decades or sometimes centuries, decisions made at the design stage, especially in the initial stage, have long term effects on the energy balance and the environment sustainability. Increasing urbanization and industrialization have degraded the environment. Uncontrolled development has important consequences on the urban climate and the environment efficiency of buildings. Higher urban temperatures have a serious effect on the electricity demand by an air conditioned of buildings and increase in smog production.

Many see the invention of mechanical heating and air-conditioning system as having liberated architects from constraint of climate. However, there is still a significant demand for natural ventilation, day lighting, and passive environmental control. The right orientation and fenestration choice can halve the air-conditioning requirement. Incorporating passive solar elements and natural ventilation pathways can eliminate it altogether.

2.2 Energy Design Tool for Architectural Design

During the conceptual stage, architects work mostly on the building plan configuration element, form, massing and orientation that are based on the given site, programmable space, zoning and building codes. Building mass, form and selection of materials are often the result of an aesthetic process, rather than rational integrated process such as environmental design and sustainability concept. Incorporating passive solar elements and natural ventilation can help energy efficiency. The building form can even be designed to provide shading using its own fabric without any need for additional structure shading. It is significantly more difficult and costly to retrofit these features at a later stage of a project's development.

If the role of the design tool is to serve the design process, then a new approach is required to accommodate the conceptual phase. The argument is that an experienced designer or architect, while capable of producing the building in this way should also receive reliable knowledge based information of the physical process involved in the building's environmental performance. Since these processes can be very complex and often highly interrelated or even counter intuitive, reliable information, tool or guideline can be very helpful in architectural decision-making. It is also the argument of this dissertation that the needs of the architects are quite different from the needs of the engineers, and that existing building design and performance analysis tools inadequately serve architects needs.

14

The conceptual stage of design occurs very early in the design process. This is the time when vast arrays of competing requirements are shaping the initial building form, when geometry, materials and orientation are still being formulated. As these are arguably the three most important determinants of building performance, this is the most crucial stage of a project.

Conceptual design is an iterative process of generating ideas that then need to be evaluated and tested, after which they are rejected or included further refinement. The main criterion for these tests is speed. Being able to quickly reject impractical ideas can save significant amounts of time, each newly rejected idea providing one more clue to a more acceptable one. A major part of this testing process is trial and error - simply playing around or experimenting with an idea until it is shown to work or not. The purpose of this is to gain some understanding, both spatially and operationally, of the full requirements of the final form. (Akin, O., 1978) Using traditional techniques, the range of testing that can be performed is quite limited.

In order to make environmental performance a practical consideration at this early stage, thereby informing the decision-making process as much as any other consideration, real and useful feedback has to be produced from what is often ill-defined and abstract information. The precise and detailed input requirements of most existing design tools preclude this. To use them, the designer must first enter the small amount of hard data they do have, and then arbitrarily quantify whatever else is needed before a result can be produced. Overcoming this requires a completely different approach from the concise, solution-based nature of existing analysis tools.

In order to be used at the earliest stages in the design of a building, any next-generation design tool must overcome the psychological separation between design and analysis that existing tools have created. Extensive quantitative input requirements in such tools act to produce a psychological separation between the act of design and the act of analysis. As discussed previously, the primary cause of this is the detailed nature and amount of input required to describe a building model. Having to enter this data very early in the design acts to interrupt the process of iterative decision-making and forces the designer to prematurely make a series of arbitrary decisions just to produce a model acceptable to the tool.

A conceptual design tool must make the process of entering this data part of the design process itself. This is only possible if there are enough tangible benefits associated with having a model in such a format. The key to this is feedback, producing real and useful design feedback at every stage of the modeling process from data entry right through to final analysis. This places the focus firmly on the interface, the means by which the user describes and interacts with the model.

At the conceptual stage, building geometry is fluid and subject to constant change, with solid quantitative information relatively scarce. Having to measure off surface areas or search out the emissivity of particular material forces the designer to think mathematically at a time when they are thinking intuitively. It is, however, at this intuitive stage that the greatest potential exists for performance efficiencies and environmental economies.

Performance characteristics can be modeled using existing design tools and those results can be used to aid building design. Special skills may still be required, but just as important is the ability to translate the roughest architectural sketch into a valid input model and translate the result into fundamentally solid design feedback.

2.3 Energy Analysis Tool for Building Simulation

Since the early 1960s, the use of computer modeling and simulation tools within the building industry has steadily increased. These tools have progressed from simple, single task applications with limited input and output requirements, (Howard, H., 1960) to quite sophisticated modeling systems that can simultaneously analyze a range of performance parameters.

Since the energy crisis in 1973, developments in computation energy analysis tools have been developed. A significant amount of the research referred to Manning, has been directed into the development of computer software for building simulation and performance analysis. A wide range of computation tools are in wide spread use in both research and commercial applications. Some focus of development in this area has long been on the accurate simulation of fundamental physical process such as mechanisms of heat flow through materials and the transmission of light. Adequate description of boundary conditions for such calculation usually requires very detailed mathematical models. To understand how air flows and heat transfers around the building, architects need a very sophisticated modeling technique. Computer modeling help solves equations that govern fluid flow, and the transferring of momentum, energy and mass. The data can be entered and translated to resolve into easy to read graphics.

Architects are mainly responsible for schematic and conceptual design of the whole building, and engineers are responsible for calculation, modeling and simulation of how to make them work. However, after nearly 30 years of development, there is still some skepticism as to the necessity and applicability of such tools in the design process. The basic question still remains: Does the use of simulation and validation tools actually produce better buildings? The main focus in the development of design tools has been the accurate simulation of natural processes, heat flow through materials, the turbulent movement of air and the inter-reflection of light. The result is a range of software tools well suited to the task of detailed design validation. However, the user interfaces and specialist skills required to properly drive them mean that they are still very much a part of the engineering domain.

There are a number of ways to calculate the non steady state of dynamic response of the building. The following are important to mention:

1. The simplest of these is the Admittance method as described in Section A9 of

   volume A - Design Data, The Chartered Institute of Building Services

Engineers 1986 Guide. This is based on a steady state calculation but
simulates the dynamic performance as fluctuations about a mean.

2. More advanced methods such as the American Society of Heating
   Refrigeration Air-condition Engineers response factor and finite element
   difference methods are more accurately represent the dynamic response of
   buildings. They are more computationally intensive and require a more
   precise building model. Simulation tools based on these methods such as
   DOE-2 (energy performance, design, retrofit, research, residential and
   commercial buildings); ESP(r) (energy simulation, environmental
   performance, commercial buildings, residential buildings, visualization,
   complex buildings and systems) and ENERGY PLUS (energy simulation,
   load calculation, building performance, simulation, energy performance, heat
   balance, mass balance) are highly recognized and have undergone a number of
   validation processes.

3. Difference Method more accurately represents the dynamic response of
   buildings. These are significantly more computationally intensive and require
   a very precise building model. Simulation tools based on these methods, such
   as AIRPAK (airflow modeling, contaminant transport, room air distribution,
   temperature and humidity distribution, thermal comfort, computational fluid
   dynamics (CFD)), DOE-2. Energy Plus and ESP(r).

By accurately modeling the physical processes of heat and airflow, it is possible
to simulate more complex thermal systems such as under-floor heating, chilled beams,
displacement ventilation, passive solar elements and natural ventilation systems. Such

tools can also calculate sensible and latent loads, radiant temperatures, inter-zonal exchange and internal solar gains tracked through multiple zones. It is possible to simulate complex thermal systems and calculate sensible and latent loads, radiant temperatures, internal zonal exchange and internal solar gains tracking through the number of zones. Such methods are known as steady state calculations as there is no accounting for the thermal response of the building fabric to cyclical fluctuation in temperature. There are a number of methods of calculating the non-steady state or dynamic response of a building.

Computer analysis tools can provide valuable results at the beginning during schematic design phase of a project, and continuing through value engineering activities while project is developed. Many of the software tools employed in new building design, energy use prediction and energy audit calculations are time consuming and often costly to use. Hourly building energy simulation models such as DOE-2, BLAST (Energy performance, design, retrofit, research, residential and commercial buildings), and Bunyip have proved to be an effective method of simulating energy use in new building during the design stage, and such models are increasingly being used to evaluate retrofits in existing buildings. DOE-2 calculates the hourly energy use and energy cost of a commercial or residential building given information about the building's climate, construction, operation, utility rate schedule and heating, ventilating, and air-conditioning (HVAC) equipment.

The activities were mainly driven by the need to upgrade the existing local energy standards to incorporate with the energy performance. Such models are increasingly being used to evaluate retrofits in existing buildings and for purpose of demand-side management (DSM) evaluations. Heat Balance Method and the Weighting Factor Methods described in ASHRAE 1981, yield the complex set of information of the building performance. Subsequent computer based algorithms include ASEAM-2, HVACSIM +(HVAC equipment, systems, controls, EMCS, complex systems), E20-II and Trace (Trane Inc, 1987).

It has been shown by Curtis (Curtis et al, 1984) that through computer simulations, reasonable accurate predictions of building energy use can be made against actual measurements of existing building types in the United States. An alternative method is to create a database of energy use, by performing a large number of computer simulations for several important building parameters, and then use the database to develop a simplified energy estimating methodology. They have created a database of energy use, by performing a large number of computer simulations for several important building parameters, and then use the database to develop a simplified energy estimating methodology. Woods in 1982 used stepwise regression and analysis of covariance to establish predictor variables of energy. Data can be gathered from simulation or from real survey data collections.

In "the effects of multi-parameter changes on energy use of large buildings" by Chou , S.K., W.L.Chang, Y.W.Wong, 1993, the author describes the development of a methodology to predict the effects of multi-parameter changes on the energy uses of large buildings. The methodology is based on the third order of Taylor's series expansion whose coefficient is evaluated for several key building parameters. The building was coded for energy performance simulation by the DOE-2 computer program. A database of simulation results was created from which the coefficients of the Taylor series expansion were derived.

IssaacTuriel (Turiel, I., Richard B., Mark S., and Mark L., 1984) have used DOE-2 simulations to develop a simplified energy estimating equation for a generic building in Denver, Colorado. The data used consists of a single parameter and simultaneous two parameters change, which have the most significant effect on building energy consumption. Through the use of curve fitting for a single parameter simulation results, and the Taylor's series expansion for two parameters results, they showed that it was possible to make predictions of building energy performance using simplified methodology. Tureil wrote that in their parametric energy analysis performed for their sixteen building parameters with three HVAC system types. Five of these variables:

orientation, ground reflectance, window setback ratio, roof and wall absorption have very small (< 2% change) total energy use impacts in Denver's climate region.

Chou and Wong (Chou, S.K. and Wong, Y.W., 1986) predict annual total energy use and cooling energy of a generic office building, conducted a similar study in Singapore. In their effort to audit energy of large buildings, they have found that building designers and engineers often need to understand the effects of multi-parameter changes on energy performance of a building. This is often critical when engineers have to make retrofitting evaluation and try to comply with local energy standards.

2.4 Justification of Using DOE-2 for Processor

Lawrence Berkeley National Laboratory, Hirsch & Associates, Consultants Computation Bureau, Los Alamos National Laboratory, Argonne National Laboratory and University of Paris developed DOE-2. The U.S. Department of Energy; the Gas Research Institute, Pacific Gas & Electric Company, Southern California Edison Company, Electric Power Research Institute, California Energy Commission provided major support and others provided additional support. Because it is scientifically rigorous and open to inspection, DOE-2 has been chosen to develop state, national, federal, and international building energy efficiency standards, including:

- The ASHRAE-90.1 standard for commercial buildings, which is based on thousands of DOE-2 analyses for different building types and climates, is mandatory for new federal buildings, and has been adopted by many states for non-federal buildings.
- The ASHRAE-90.2 standard for residential buildings, is based on 10,000 DOE-2 analyses.
- The State of California standard for commercial buildings (Title 24).
- Standards for other countries, such as Hong Kong, Saudi Arabia, Kuwait, Singapore, Malaysia, Philippines, Indonesia, Thailand, Switzerland, Brazil, Canada, Mexico and Australia.

DOE-2 is hourly, whole-building energy analysis program calculating energy performance and life cycle cost of operation. Can be used to analyze energy efficiency of given designs or efficiency of new technologies. Other uses include utility demand-side management and rebate programs, development and implementation of energy efficiency standards and compliance certification, and training new corps of energy-efficiency conscious building professionals in architecture and engineering schools.

DOE-2 is an up-to-date, unbiased computer program that predicts the hourly energy use and energy cost of a building given hourly weather information and a description of the building and its HVAC equipment and utility rate structure. Using DOE-2, designers can determine the choice of building parameters that improve energy efficiency while maintaining thermal comfort and cost-effectiveness. The purpose of DOE-2 is to aid in the analysis of energy usage in buildings; it is not intended to be the sole source of information relied upon for the design of buildings: The judgment and experience of the architect/engineer still remain the most important elements of building design.

The Figure 2-1 shows a flowchart of DOE-2. Basically, DOE-2 has one subprogram for translation of your input (BDL Processor), and four simulation subprograms (LOADS, SYSTEMS, PLANT and ECON). LOADS, SYSTEMS and PLANT are executed in sequence, with the output of LOADS becoming the input of SYSTEMS, etc. The output then becomes the input to ECON. Each of the simulation subprograms also produces printed reports of the results of its calculations.

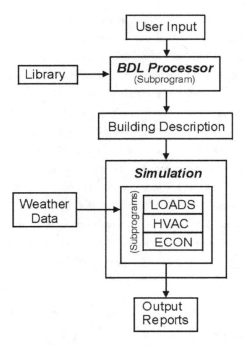

Figure 2- 1Element of DOE-2

DOE-2 is the source of algorithms, calculation techniques, and correlations for many widely used simplified methods. These include:

- ASEAM-2 simplified energy analysis program
- ADM-2 simplified energy analysis program
- TrakLoad and LoadShaper simplified energy analysis programs
- RESEM program for retrofit analysis
- EEDO (Energy Economics of Design Options)
- Daylighting Nomographs
- Energy Nomographs

- AAMA-SKY program for skylight design
- ENVSTD program for ASHRAE Standard 90.1
- PEAR program for residential analysis
- RESFEN (Residential Fenestration Performance Design Tool)
- COMFEN (Commercial Fenestration Performance Design Tool)

One of the ways to reduce time and cost of performing large building energy analysis is to write a faster running program with simplified algorithms that can be reasonable and accuracy and flexibility in use by Kusuda and Sud, used an early approach in *1982*, in developing the modified bin method for energy prediction. Subsequent computer based algorithms include ASEAM-2 (Fireovid and Fryer, 1987), HVACSIM + (Clark and May, 1985), E20-II (Carrier Corporation, 1988), and Trace(Trane Inc, 1987). The private sector has adapted DOE-2 by adding interfaces that make the program easier to use. Some examples are shown in Table 2-1. DOE-2 has been incorporated in commercial building design software environments such as COMBINE (European Community) and RIUSKA (Finland). DOE2 results on the energy-efficiency potential of different building types have been incorporated by Pacific Northwest National Laboratory in the U.S. Energy Information Administration's National Energy Modeling System for predicting future energy demand.

Professional societies and industry groups for research, development, and impact analysis use DOE-2. For example, ASHRAE used DOE-2 for standards development and the Gas Research Institute (GRI) used DOE-2 to assess the energy economics of, and thereby determine future R+D and marketing efforts for new gas technologies, including gas-engine-driven chillers, desiccant cooling systems, direct-fired absorption cooling, and cogeneration.

Many utility companies use DOE-2 as a key element in their demand-side management programs to encourage energy-efficiency as an alternative to building new power plants. For example: Northeast Utilities, Pacific Gas & Electric and Southern California Edison offer DOE-2 analysis to architects and engineers as an incentive to designing energy-efficient buildings. The Bonneville Power Authority (BPA) in its Energy Edge program used DOE-2 to show the practicality of buildings that use 30% less energy than its existing standard. Pacific Gas and Electric (PG&E), the largest investor-owned utility in the U.S., used DOE-2 in its ACT2 (Advanced Customer Technology Test) project to select advanced energy efficiency retrofit measures in residential and commercial buildings. Many States use DOE-2 to determine the potential for energy savings. For example, New York State used DOE-2 to show that adoption of cost-effective conservation measures would reduce statewide electricity consumption by 38%. States and the federal government use DOE-2 to forecast the long-range cost and energy savings of building energy efficiency programs. The National Fenestration Council (NFRC) has used DOE-2 to develop window energy efficiency labels.

Because of its accuracy, DOE-2 is used as a reference standard program. Two examples of this are:

1. ASHRAE validated its widely used simplified energy calculation method (the TC 4.7 bin method) by comparing its results with DOE-2. This comparison also led to improvements to the TC 4.7 method.

2. The California Energy Commission certifies computer programs for use in Title 24 compliance by requiring that they agree with DOE-2 to within a certain percentage on a set of test buildings.

DOE-2 is used in 60+ universities in the U.S. for building science research and for teaching. National labs, universities, and industry for hundreds of studies of products and strategies for energy efficiency and electric demand limiting have used DOE-2. Examples include advanced insulating materials, evaporative cooling, low-E windows, switch able glazing, day lighting, desiccant cooling, cogeneration, gas-engine-driven cooling, cool storage, effect of increased ventilation, sizing of thermal energy storage systems, gas heat pumps, thermal bridges, thermal mass, variable exterior solar and IR absorptance, and window-performance-labeling.

DOE-2 has undergone validation by Los Alamos National Laboratory, Lawrence Berkeley National laboratory and universities to show that that the program can accurately predict energy use in real buildings. Such validation gives users confidence that the DOE-2 results are reliable for well-described buildings.

2.5 Propose New Innovative Design Tool for Architects

This research attempts to alleviate problems that architects have during their initial stage of design by introducing the new concept and method to make this tool applicable to any buildings. That is why proposing the creation of representative module, which is suitable for the schematic design stage and more response to climate and environment. It is important for sustainable architecture to embrace this new software tool incorporated into the architectural design and building delivery process. The climatic data used in simulation can influence and address the potential of environmental energy forms and their changing levels through out the year. Building envelope especially thermal control and daylight encounter tremendous environmental impact. The provision of daylight in a building is strongly linked to the spatial and architectural design, while thermal control is linked to the comfort and well being of the occupants. How we can create buildings' space that human being has to work, use, and perform with positive and productive consequences while integrating the built structure with ecological and bio-climatic system.

CHAPTER 3

DEFINE MODULE AND ITS PARAMETERS

Having the module representing the building makes the study of energy efficiency become relative importance by comparing between each design schemes, rather than getting the absolute numbers. There is an importance definition of the significance difference between study of the relative comparison and the absolute one. Relative comparisons are used frequently when we want to know which design schemes or methods are more energy efficient. In the architectural preliminary stage, architects do not need real precise absolute answers, although results from DOE2.1E simulation gave very accurate results.

3.1 Module as a Conceptual Model

The module is intended to represent a hypothetical concept design study in representing idea, and demonstrate the method, program and process, which led to the new learning methods. The program and process created in this research can be used for any building type, function, and climate location. Four Methods have been developed:

1. Customized model

2. Interactive and interfacing model program or front end software

3. Library entry data

4. Programming the model

The DOE-2.1E building energy simulation program engine was used for a model in order to generate data for the study and permit the development of new learning methods. The main advantage of this methodology is that it enables an architect without specialized computer skills to perform energy analysis and prediction. Thus energy simulation model and synthesis can be done more quickly and with less investment.

Simulation is the process of developing a simplified model of a complex energy usage in a building, and using this model to analyze and predict the behavior of the energy performance. This research uses the energy modeling technique, based on understanding that a building is a complex network of thermal resistance and capacitance linking different regions and representing conduction, convection, radiation, and heat storage processes. The manner, in which this network is treated mathematically, while simplifying boundary condition assumption that might be made, determines the flexibility of the modeling technique to emerge.

The simulation model, analysis, synthesis, and results will reveal prediction problems and energy opportunities that the designer can consider. The guidelines and recommendations can provide the architect with visual qualitative and quantitative information about how the building reacts and various architectural design parameters interact when the design scheme is altered. The results can be used to compare the energy design performance tradeoffs in a number of options. This research also explains the precise relationship between the output results and the design input parameter analytically and explicitly, taking into account the complex effect of coupling by the building load.

The module has three enclosed walls-- one side has a window that faces exterior climate conditions; the other two sides are solid wall. The inner space has one interior partition, floor and ceiling. The environmental exchange occurs only through the window facing the assigned orientation.

This hypothetical module as shown in Figure 3-1 can represent any mid space or mid floor, corner space in a typical building in the site and climate location. The building schedule used the typical office-building schedule from e-Quest and DOE-2.2. Module wall, floor, partition, and ceiling used the typical office building construction with insulation. The intent of the research model is to see the impact of the parameters' interaction. An assumption was made that the module complies with the code practice as detailed below:

28

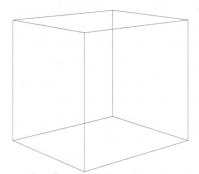

Figure 3- 1 Picture Idea for the Module

3.2 Module Geometry

Module as shown in Figure 3-2, Figure 3-3, and Figure 3-4 has the size of 15 feet x 15 feet, 14 feet floor-to-floor height, and 10 feet floor-to-ceiling height. Module has 5 feet perimeter depth with the capability to link or relate data as well as store data. It is important to study the corresponding effects of the perturbation in the module in order to understand the relative importance of the design input parameters and their impact on the output results.

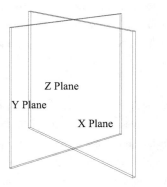

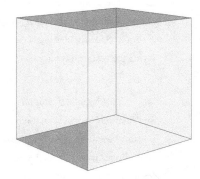

Figure 3- 2 Module Geometries

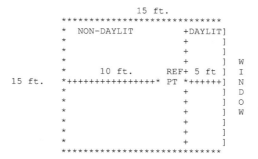

Figure 3- 3 Module Plan-Perimeter Depths

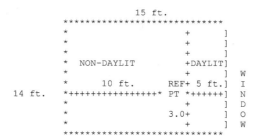

Figure 3- 4 Module Section

Module has the features that compliance with code and standards

- Module area equals to 225sq.ft.
- Module volume equals to 2025cu.ft.
- Lighting load is 2.0 watt / sq.ft.
- Light heat transfer to space equals to 0.80
- There are no people and equipments presented in the module.
- Typical wall construction has U-value equals to 0.08 Btu/h-ft-°F.
- Typical partition construction has U-value equals to 0.5 Btu/h-ft-°F.
- There is no heat transfer between the partitions.

- Indoor design condition temperature is 72°F for both summer and winter.

The U.S. Department of Energy (DOE) is working to improve the energy efficiency of the Nation's buildings through new technologies and better building practices. Energy-efficient buildings improve the lives of Americans by saving consumers money, lessening our demand for fossil fuels, decreasing the need for new power generation, and reducing environmental emissions. DOE's Building Energy Codes Program is an information resource on national model energy codes.

Strengthening energy codes increases the likelihood of energy and cost savings in new construction and renovations to existing buildings. New buildings can be designed to be both more comfortable and more efficient, cutting heating and cooling costs by close to 50%. The most cost-effective time to improve the envelope, equipment, and lighting efficiency of existing buildings is when those elements must be replaced or upgraded for new owners or tenants or due to normal wear and tear.

Department of Energy works closely with the International Code Council (ICC), the American Society of Heating, Refrigerating and Air-Conditioning Engineers, Inc. (ASHRAE), the Illuminating Engineering Society of North America, and other code user groups to develop more stringent and easier-to-understand building energy codes. These groups are identifying candidate energy-efficient technologies and practices and working to remove barriers to these technologies in the national model energy codes.

In Article13 Chapter 18-13 of the proposed energy code, City of Chicago, Table 3-1 contains the guidelines for design criteria and thermal design parameters of the exterior design condition, and shall be used for calculation. Table 3-2 contains Solar Heat Gain Coefficients for Fenestration. Table 3-3 contains U-Factor for Windows and glazing.

Table 3- 1.Exterior Design Condition 18-13-302.1

Condition	Value
Winter, Design Dry Bulb (°F)	-10°F
Summer, Design Dry Bulb (°F)	92°F
Summer, Design Wet Bulb (°F)	74°F
Degree Days Heating	6151 HDD
Degree Days Cooling	1015 CDD

The Chapter also described the building material, and envelope insulation:

Table 3- 2 SHGC Default Table for Fenestration 18-13-304.5.2(3)

Product Description	Single Glazed				Double Glazed			
	Clear	Bronze	Green	Gray	Clear + Clear	Bronze + Clear	Green + Clear	Gray + Clear
Metal Frames								
Operable								
Fixed	0.75	0.64	0.62	0.61	0.66	0.55	0.53	0.52
	0.78	0.67	0.65	0.64	0.68	0.57	0.55	0.54
Non-Metal Frames								
Operable								
Fixed	0.63	0.54	0.53	0.52	0.55	0.46	0.45	0.44
	0.75	0.64	0.62	0.61	0.66	0.54	0.53	0.52

Table 3- 3 U-Factor Default Table for Windows 18-13-304.5.2(1)

Frame Material and Product Type	Single Glazed	Double Glazed
Metal without Thermal break		
Operable (including sliding and swinging glass door)	1.27	0.87
Fixed	1.13	0.69
Garden Window	2.60	1.81
Curtain Wall	1.22	0.79
Skylight	1.98	1.31
Site assembled sloped, overhead glazing	1.36	0.82
Metal with Thermal break		
Operable (including sliding and swinging glass door)	1.08	0.65
Fixed	1.07	0.63
Curtain Wall	1.11	0.68
Skylight	1.89	1.11
Site assembled sloped, overhead glazing	1.25	0.70
Reinforced vinyl, Metal clad wood		
Operable (including sliding and swinging glass door)	0.90	0.57
Fixed	0.98	0.56
Skylight	1.75	1.05
Wood, Vinyl, Fiberglass		
Operable (including sliding and swinging glass door)	0.89	0.55
Fixed	0.98	0.56
Garden Window	2.31	1.61
Skylight	1.47	0.84

* Glass block assembles with mortar but without reinforcing or framing shall have a U-Factor of 0.60

The investigations of design performance parameters range that lead to sophisticated modeling and simulation tools as part of new approach in the design process are necessary. A wide range of values was studied for each parameters even this range might not reflect the current possibilities that found in the actual buildings.

The relationships of these variables and parameters are complicated and need more simplified simulation system and quantify energy model to handle. The essence of this research is to simplify the energy model and simulation through understanding of the method, process and the practical implication, which will assist the designers overcoming these problems.

The following examples are some parameters that represent the energy conscious measures. Some parameters in the module are variables some are constant. The Followings are the variable design parameters

3.3 Climate Change

Chicago is in between cold zone and temperate climate, which encountered cold and conservation of heat as well as reduction of excessive heat impact and provide shade in the summer. Climate location variables that were built into the software are the followings:

- Climate Location: Chicago
- Latitude = 42
- Longitude = 88
- Altitude = 610
- Time-Zone = 6

Bangkok is in hot and humid climate, which presented the avoidance of excessive solar radiation and the evaporation of moisture by breeze. The built environment needs to allow free air movement and bring shading into effect.

- Climate Location Bangkok
- Latitude = 13.92

34

- Longitude = -100.60
- Altitude = 39
- Time-Zone = -7

Figure 3- 5 Climate location: Chicago (Red circle left) and Bangkok (Red circle right)

Figure 3-5 shows the climate location of both cities. The weather data for a location consists of hourly values of outside dry-bulb temperature, wet-bulb temperature, atmospheric pressure, wind speed and direction, cloud cover, and (in some cases) solar radiation. Figure 3-6 shows the weather map in April 28, 2005 at 5AM USA Central Time.

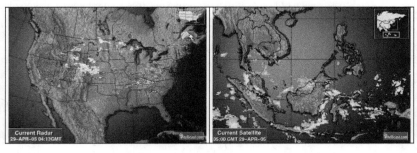

Figure 3- 6.Weather Map in April 28, 05@ 5AM, Chicago (left) and Bangkok (right)

Running the DOE-2 weather processor on raw weather files provided by the U.S. National Weather Service and other organizations produces weather data suitable for use in DOE-2

Design conditions contained here for Chicago and Bangkok, provide information on station location, which represents the temperate and tropical climatic zone region, period analyzed, the heating design conditions, cooling design condition, humidity, wind, mean annual extreme and standard deviation of minimum and maximum dry-bulb temperature, and mean daily temperature range. For Chicago the station is O'Hare International Airport, for Bangkok the station is Don Meong International Airport.

The weather data is very important factor in the architecture envelope load simulations since the heating and cooling energy loads are the weather dependent. In DOE-2 a weather database, which is a representation of the prevailing weather condition, is required an hourly weather data. The weather data is compiled from the raw weather file that is collected from National Oceanic and Atmospheric Administration's raw weather data Test Reference Year or Typical Metrological Year (TRY or TMY) format. Then process into the DOE-2 weather processor, which is a batch command line program called doewth or doewth.exe. The primary function of the weather processor is to read hourly weather data, extract the data needed by DOE-2, and write packed binary weather file that is used by the DOE-2 simulation program. In addition to its primary function (called packing), the weather processor can produce hourly listings of raw or packed weather files in a readable format, and can produce a summary report of the data on a packed DOE-2.1E weather file.

The DOE-2 weather file contains hourly data for one year (8760 hours). All DOE-2.1E weather files are 365 days long. The hourly data on the weather files for Chicago was obtained from a Test Reference Year or Typical Metrological Year (TRY or TMY) tape. The DOE-2 weather processor can process data in this format and add modeled solar data into the DOE-2 weather file, and the DOE-2 simulation program will calculates solar values using the ASHRAE clear sky model and the clearness numbers, cloud amounts, and cloud types from DOE-2 weather file. The solar DOE-2 weather files contain the hourly values of the total horizon solar radiation (Btu/hr-ft²) and direct normal solar radiation (Btu/hr-ft²)

Test Reference Year (TRY) format is an old, typical year data set that did not include solar radiation data. The format can still be useful as a format for measured data. The DOE-2 weather processor can process data in this format and add modeled solar data into the DOE-2 weather file, and the DOE-2 simulation program will calculates solar values using the ASHRAE clear sky model and the clearness numbers, cloud amounts, and cloud types from DOE-2 weather file. The solar DOE-2 weather files contain the following hourly values;

1. Total horizon solar radiation (Btu/hr-ft²)

2. Direct normal solar radiation (Btu/hr-ft²)

Figure 3-7 shows the Bio-Climatic data representation of one of US major city.

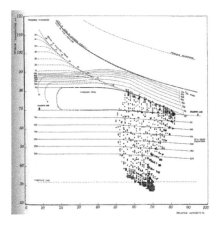

Figure 3- 7.Bio-climatic data regulate comfort (Olgyay, "Design with climate").

The components in the weather files are included the followings
1. Dry-bulb Temperature

38

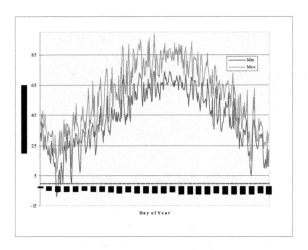

Figure 3- 8 Chicago Dry Bulb Temperature (°F) for 365 days of Test Reference Year

2. Wet-bulb Temperature

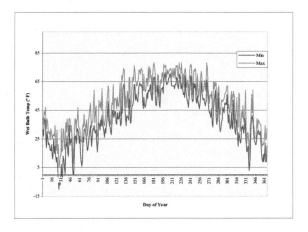

Figure 3- 9.Chicago Wet Bulb Temperature ((°F) 365 days of Test Reference Year

3. Atmosphere Pressure (inches if Hg times 100)

4. Wind speed or velocity (Knots)

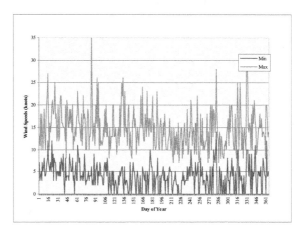

Figure 3- 10 Chicago Wind Speeds (knots) for 365 days of Test Reference Year

5. Wind direction (compass points 0-15, with 0 being north, 1 = NNE, etc.

6. Cloud amount (0 to 10, with 0=clear and 10= total overcast)

7. Cloud type (0, 1, or 2) (0 is cirrus or cirrostratus, the least opaque, 1 is stratus

   or stratus fractus, the most opaque, 2 is all other cloud types, of medium

   opacity.

8. Direct and diffuse solar radiation for 8760 hours of the year.

9. Humidity ratio (lb of water per lb of dry air)

10. Density of the air (lb/ft³)

11. Specific enthalpy of the air (Btu/lb)

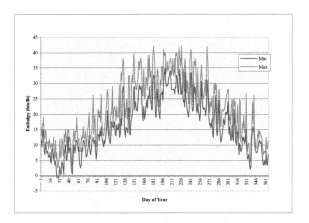

Figure 3- 11.Chicago Enthalpies (Btu/lb) for 365 days of Test Reference Year

12. Rain flag (0 means is not raining, 1 means it is)

13. Snow flag (0 means is not snowing, 1 means it is)

The followings are the examples of weather files. Many of them can be obtained for free through the Internet <www.doe2.com> [SAC Software, 1999].

1. **CTMY (Canadian TMY)** has 12 locations, derived from Canadian

   Government data.

2. **CTMY2 (Canadian TMY updated)** has 40-location version, derived from

   Canadian Government data.

3. **CZ2 (California Climate Zones Revision 2)** has 16 zones, dated in 1992,

   supplied by the California Energy Commission.

4. **TMY (Typical Meteorological Year weather file)** The TMY file contains

   the average of the hourly climatic data for 238 locations, derived from United

States National Oceanic and Atmospheric Administration's N.C.D.C. TMY datasets)

5. **TMY2 (Typical Meteorological Year version 2**) has 238 locations, derived from U.S. Department Of Energy's NREL datasets)

6. **TRY (Test Reference Year**) has 60 locations, derived from United States National Oceanic and Atmospheric Administration's NCDC datasets.

7. **WYEC (Weather Year For Energy Calculations**) has ASHRAE 51 locations, derived from United States National Oceanic and Atmospheric Administration's NCDC datasets.

8. **WYEC2 (WYEC version 2**) has 51 ASHRAE locations, derived from United States National Oceanic and Atmospheric Administration's NCDC datasets.

Table 3- 4 DOE-2 Weather Processor Test Reference Year or Typical Metrological Year (TRY or TMY) Tape

								3=smoke
								4=haze and smoke
								5=thunderstorm
								6=tornado
								7=liquid precipitation
								8=frozen precipitation
								9=blowing dust
148190080070003000082956810103025999999999999999999999999999 4845607999907324000499990059999999999999999999999999999 2000010100							1974010100	
City				Columns Number				
	01-05	06-08	09-11	12-14	15-17	18-20	21-24	25
	Station Number	Drybulb Temp	Wetbulb Temp	Dew point Temp	Wind Direction	Wind Speed	Station Pressure	Weather Type
Chicago								0=no obstruction
Bangkok								1=fog
								2=haze

City	Columns Number						
	26-27	28-29	30	31-33	34-35	36	37-39
	Total sky cover	Amt of lowest cloud layer	Type of lowest cloud	Height of base of lowest layer	Amt of second cloud layer	Type of cloud second layer	Height of base second layer
Chicago							
Bangkok							

City	Columns Number						
	40-41	42-43	44	45-47	48-49	50-51	52
	Sum of first two layer	Amt of third cloud layer	Type of cloud third layer	Height of base third layer	Sum of first three layer	Amt of fourth cloud layer	Type of cloud fourth layer
Chicago							
Bangkok							

City	Columns Number							
	53-55	56-59	60-69	70-73	74-75	76-77	78-79	80
	Height of base fourth layer	Solar radiation	Blank: reserve for future use	Year	Month	Day	Hour	Blank: reserve for future use
Chicago								
Bangkok								

3.4 Site Orientation

Site orientation and azimuth location describes the energy uses as the function of the climate and location. Arrangement of the buildings in latitude on the site from North-

South has significant influenced on heating and cooling load, from the solar heat source stand point, different from the building arranged in longitude from East-West. Figure 3-12 shows the charts for determine the direct and diffuse solar energy, unit of Btu/ft2-hr, on vertical surface. Figure 3-13.shows the average daily solar energy received in January and in July. Site size The land area also play an important role in determining the overall shape of the building, which when combined with building orientation will influence the potential heat transfer through the building. Heat gains through envelope all 4 sides all year round. North facing direction is subjected to minimum heat gain in summer associated with maximum heat loss in winter. Orientation variables that built into the software describe how's the module's façade facing solar radiation, which are North, Northeast, East, Southeast, South, Southwest, West, Northwest. Figure 3-14 shows the module's façade faces different orientation.

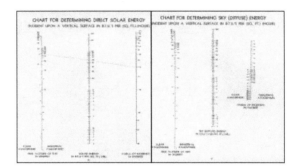

Figure 3- 12.Charts for determine the direct solar energy (Left) and diffuse solar

energy (Right) Btu/ft2-hr on vertical surface (Olygay, "Design With Climate")

44

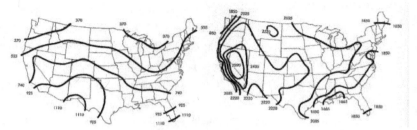

Figure 3- 13.Average Daily Solar Energy Received in January (Left) and in July

(Right)In USA (Olygay, "Design With Climate")

North   Northeast   East   Southeast   South   Southwest   West   Northwest

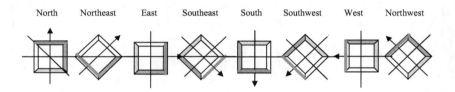

Figure 3- 14.Module's façade faces different orientation

3.5 Window Wall Facade Ratio

Variation in building shape greatly affects the amount of exterior surface area for a given volume enclosed. Skin glazing will determine in large part of the amount of energy the building will use, if a significant portion of heating and cooling load is due to heat flow through. The building form determines the building surface area to volume ratio, which in turn affects the amount of solar gain on the envelope and the potential amount of heat transfer into and out of the building through envelope. Heat is a form of energy, which appears as molecular motion in substances or as radiation space. Heat is transferred through a building envelope by conduction, radiation, or convection. Conduction is a process of heat transfer in a solid or fluid at rest by direct molecular interaction. Convection is the process of heat transfer by flowing and mixing motions in fluids. Radiation is the process of heat transfer by means of electromagnetic waves. Heat flows (gain or loss) in a building are affected by solar radiation, internal gains and temperature differences, thus created energy demand.

Building form will determine in large part of the amount of energy the building will use, if a significant portion of heating and cooling load is due to heat flow through the envelope. Building envelope exposure or the skin facades shown in Figure 3-15 has the window to wall surface area ratio. Window surface area ratio has the significant impact on the energy performance of the building. Window has solar transmission, glass conductance (Btu/hr-ft²ºF) and shading coefficient, according to its variables that changes. The criteria include evaluating the followings to compare energy savings, and natural lighting, according to the variables.

- Window (or curtain wall) performance
- Glazing performance
- Shading-coefficient
- Glass conductance
- Window setback
- Lighting power
- Heat-of-lighting to space ratio

Figure 3- 15.Skin facades (Victor Olygay, "Design With Climate")

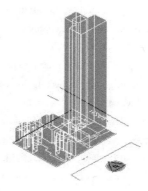

Figure 3- 16. Building forms provide variation of building surface area

Figure 3-16 shows the building forms that provide variation of building surface area Envelope and fenestration represented by an arrangement of building construction materials with specific thermal transmittance or conductance including opaque wall above grade, ceiling, roof, slabs, floors, glazing, or windows. Glass curtain wall provides natural lighting to interior spaces. The amount of glass area becomes the important issues to moderate the energy aspects of electric lighting and natural lighting, heating and cooling, and the need for visual views. Different ratio of glazing in external walls for each orientation change the overall building performance, and impact form solar transmittance and shading coefficient of the glazing.

Fenestration includes the light transmitting envelope component assemblies in a building wall, building materials and cladding, Color and texture of the exterior wall facing, U value of the opaque wall and the solar absorbance of the exterior wall. Thermal mass is the wall (or roof) material with yields significant heat capacity and surface area that affects building loads by absorbing or releasing heat due to the fluctuation.

The envelope gain or loss depends on three variables:

1. Temperature difference

2. Exposed area refers to number of the square feet of surface exposed to the
    temperature difference. As with solar computations, exposed area is a function
    of geometry and can be manipulated. Increasing exposed area will increase
    thermal transmission. Changing geometry can alter the amount of exposed
    area.

3. Transmission: changing materials and construction detailing of the envelope
    can alter transmission. The materials and construction detailing of exterior
    envelope impede the flow of heat between exterior temperatures an interior
    temperature differences. Figure 3-17.shows heat transfer through radiation,
    conduction, and convection. Figure 3-18. shows heat is transferred through
    radiation from warm to cold area.

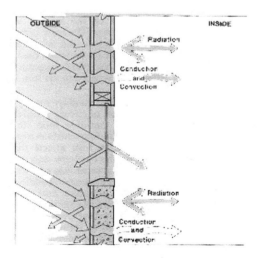

Figure 3- 17. Heat transfer through radiation, conduction, and convection

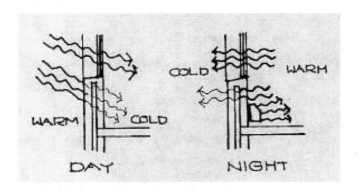

Figure 3- 18 Heat is transferred through radiation from warm to cold area

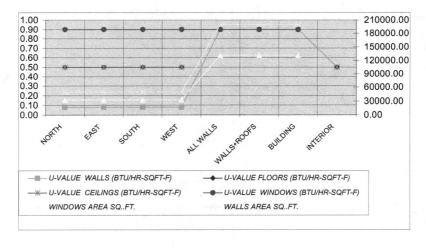

Figure 3- 19.Thermal transmissions are measured in term of U values

The most basic method of heat load analysis is based on the Building Loss Coefficient (BLC), which is a function of the U-Value and surface area of each element in the building fabric. Figure 3-19 shows thermal transmissions are measured in term of U values. This figure quantifies the amount of heat loss over the entire building for a given temperature difference between inside and outside. Climate data can then be used determine incident solar radiation on surfaces and then calculate heating and cooling loads required if some static internal conditions are to be maintained. Thermal transmission is measured in term of U values. A number of different U values for layers of envelope materials are used more in a wall than other structure part of the high-rise building. Figure 3-20 shows multiple U values for layered mass wall.

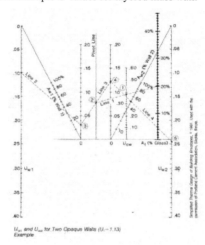

Figure 3- 20 Multiple U Values for Layered Wall (Olygay, "Design With Climate")

Fenestration variables that are built into the module are 10% of glass, 40% of glass, 50% of glass, 60% of glass, 80% of glass. Figure 3-21 shows the analysis of the window to wall ratio used in the module. Table 3-5.contains the analysis of the window to wall ratio.

50

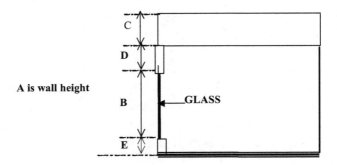

Figure 3- 21..Analysis of the Window to Wall Ratio Used in the Module

Table 3- 5 Analysis of the Window to Wall Ratio Used in the Module

Window to wall ratio	Wall height	Window height	Duct space	Window sill	Distance above glass
	A	B	C	E	D
10% glass	14	1.40	3	1	8.60
30% glass	14	4.20	3	1	5.8
40% glass	14	5.60	3	1	3.8
50% glass	14	7.00	3	1	3.0
60% glass	14	8.40	3	1	1.6
70% glass	14	9.80	3	1	0.2
75% glass	14	10.50	3	0	0.5
80% glass	14	11.20	2.80	0	0

Window height=Wall height*Window to wall ratio (varied)
Window Y coordinate (varied)

3.6 Glass Type

Modeling techniques extend from single glazing to double-glazing to triple-glazing units can now be theoretically designed for particular solar control, low emissivity and visible transmission performance. There are several ways of controlling window-shading devices such as shading coefficient and glass conductance, when solar gain, outside temperature, or daylight glare exceed the threshold value-we called these the **threshold control**.

Modeling is particularly useful in predicting the properties of coating and base glass tint combinations, and the optical and thermal properties of single, double, and triple-glazing units. A database containing of the optical properties of the Group's coated glasses and body tint glasses has been developed. Figure 3-22 shows heat performance in glass types By computer manipulation it is possible to 'lift' the coating from one base glass and place it on a second. The optical properties of novel combinations of coatings and base glass tints can be calculated without having to make test samples. This process can be taken further: the glasses can be combined into double or triple glazing units, again by computer modeling, and their properties calculated.

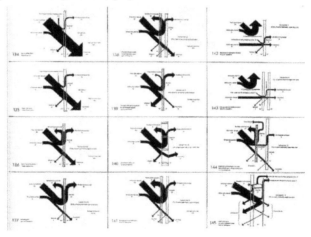

Figure 3- 22.Heat Performance in Glass Types (Olygay, "Design With Climate")

Intensive research over several decades has equipped the float process with ingenious ways of controlling solar energy for greater comfort and economy. Figure 3-23

52

shows the solar energy transmittance and wavelength for float glass types Ingredients can readily be added to the melted to control the wavelength and percentage of radiation transmitted or absorbed by the finished product. Cobalt and nickel, ferrous and ferric iron, cerium and titanium are all used to control transmission of infrared or ultraviolet waves to differing degrees and with differing visual effects.

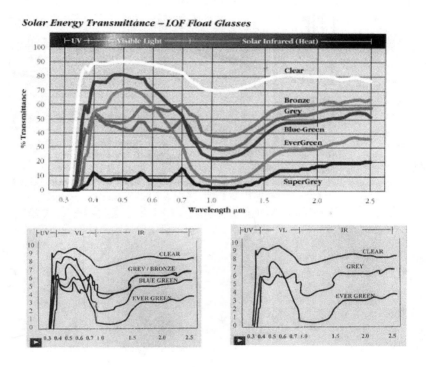

Figure 3- 23.Solar Energy Transmittance and Wavelength for Float Glass Types

Window consists of 14,3/4,1/2 inch, from single glazed clear to insulated and low-e glass curtain wall system. Consider the main wall as an aluminum frame with window units set inside the frame. The average U-Value for this wall is 0.55

BTUH/SQFT-°F. Glazing types variables that are built in to the software are the single glazed –clear, single glazed –tinted, insulated –clear, insulated –tinted, insulated –low E (low emissivity) coating clear, insulated –low E coating tinted, insulated –low E coating green, insulated –low E coating reflective.

Coatings on the glass are another way of modifying the solar control properties. Multi-layer coatings using different materials are needed for the highest optical performance. Computer modeling now means the optical properties of coatings containing seven or more layers can be predicted, with only minimal experimental trials needed before full-scale production. High performance solar control coatings can now be produced, based on ultra-thin metallic silver layers sandwiched between oxide layers, which combine high near-infrared heat reflection with high visible transmission. Table 3-6. contains glass types' properties used in the module.

Table 3- 6 Glass Types' Properties Used in the Module

Glass Type	Pane	Thickness	Air space	Shading Coefficient	Visible Transmission	U-value conductance Btu/hr/sq-ft/F
Single Clear	1	1/4 inch.(6.35 mm.)	None	0.95	0.80	1.09
Insulated Clear	2	1/4 inch.(6.35 mm.)	1/2 inch (12.7mm.)	0.81	0.61	0.55
Insulated low-E Clear	2	1/4 inch.(6.35 mm.)	1/2 inch (12.7mm.)	0.76	0.52	0.36
Insulated low-E Tinted (Green)	2	1/4 inch.(6.35 mm.)	1/2 inch (12.7mm.)	0.52	0.34	0.36
Insulated low-E Reflective	2	1/4 inch.(6.35 mm.)	1/2 inch (12.7mm.)	0.39	0.24	0.28

Code	Description	U-value	SC	Tvis
1000	Single Clear	1.11	1.00	0.90
1205	Single Tint Grey	1.09	0.69	0.43
1000	Double Clear	0.57	0.88	0.81
1205	Double Tint Grey	0.56	0.55	0.38
1000	Double Low-E Clear	0.50	0.84	0.77
1205	Double Low-E Tint	0.29	0.33	0.41

Low-Emissivity (Low-E) products, including the off-line coated thin silver-based product Pilkington **Optitherm**™, and the on-line manufactured Pilkington **K Glass**™ and Energy Advantage are based on a semi-conducting coating of tin oxide doped with fluorine atoms. The coating, 0.3-0.4 micron thick, has the unusual property of transmitting almost all-visible light, while rejecting heat waves (infrared) at room temperature. It is deposited in a few seconds by a chemical vapor deposition (CVD) process applied to the hot glass while it moves through the manufacturing process. The coating is hard, coherent and adheres well to the glass during cutting and handling, because it is deposited at high temperature. A low-E coating is normally used on the cavity-facing surface of the inner pane of a double-glazing unit. Its job is significantly to reduce radiative heat loss into the cavity and then to the outside air. Float glass itself strongly absorbs infrared and warms up, but the low-E coating radiates poorly into the cavity. Figure 3-24 shows the example of Pilkington glass manufacturing data.

Insulating Glass Performance Data[1]
[Insulating units constructed of equal glass thicknesses and 1/2" (12mm) airspace]

Product	Nominal Glass Thickness		Visible Light		Total Solar Energy		UV[1]	U-Value[1]		European U-Value (K-Value)[1]	Solar Heat Gain Coefficient[1]	Shading Coefficient[1]
	in	mm	Trans-mittance %	Reflect-ance %	Trans-mittance %	Reflect-ance %	Trans-mittance %	Summer	Winter			

Uncoated Float Glass Outer Lite and Clear Float Glass Inner Lite

	3/32	2.5	82	15	74	14	60	0.55	0.49	2.8	0.78	0.91
	1/8	3	81	15	71	13	56	0.55	0.49	2.8	0.76	0.89
Clear Float	5/32	4	80	15	67	13	52	0.55	0.49	2.8	0.74	0.86
	3/16	5	79	15	65	12	49	0.55	0.49	2.8	0.72	0.84
	1/4	6	78	15	61	12	46	0.55	0.48	2.8	0.70	0.81
	1/8	3	55	9	49	8	50	0.57	0.49	2.8	0.57	0.67
Grey Tint	3/16	5	45	7	38	7	22	0.57	0.49	2.8	0.49	0.56
	1/4	6	39	7	33	6	18	0.57	0.48	2.8	0.44	0.51
	1/8	3	62	10	55	9	31	0.57	0.49	2.8	0.62	0.73
Bronze Tint	5/16	5	53	9	45	8	23	0.57	0.49	2.8	0.55	0.64
	1/4	6	48	8	40	7	19	0.57	0.48	2.8	0.50	0.59
Blue-Green Tint	1/4	6	67	12	46	8	20	0.57	0.48	2.8	0.50	0.58
	1/8	3	69	12	42	8	23	0.57	0.49	2.8	0.51	0.59
EverGreen High-Performance Tint	3/16	5	65	11	36	7	17	0.57	0.49	2.8	0.45	0.53
	1/4	6	59	10	28	6	12	0.57	0.48	2.8	0.39	0.45
	1/8	3	22	5	20	5	6	0.59	0.49	2.8	0.32	0.37
SuperGrey High-Performance Tint	3/16	5	11	4	9	4	2	0.59	0.49	2.8	0.23	0.26
	1/4	6	7	4	6	4	1	0.59	0.48	2.8	0.20	0.23

ECLIPSE® Reflective Glass Outer Lite and Clear Float Glass Inner Lite

ECLIPSE Clear (1)*	1/4	6	56	16	48	32	8	0.55	0.48	2.8	0.46	0.53
ECLIPSE Clear (2)*	1/4	6	56	42	56	27	6	0.56	0.48	2.8	0.47	0.55
ECLIPSE Blue-Green (1)*	1/4	6	30	46	25	30	5	0.56	0.48	2.8	0.51	0.56
ECLIPSE Blue-Green (2)*	1/4	6	30	33	23	17	5	0.57	0.48	2.8	0.33	0.39
ECLIPSE Grey (1)*	1/4	6	17	45	20	31	3	0.56	0.48	2.8	0.29	0.34
ECLIPSE Grey (2)*	1/4	6	17	13	21	11	3	0.57	0.48	2.8	0.32	0.34
ECLIPSE Bronze (1)*	1/4	6	21	46	24	32	3	0.56	0.48	2.8	0.32	0.38
ECLIPSE Bronze (2)*	1/4	6	21	19	24	14	3	0.57	0.49	2.8	0.35	0.41
ECLIPSE Gold (2)*	1/4	6	29	15	26	32	2	0.56	0.48	2.8	0.35	0.40

*Number in parenthesis indicates surface where coating is located.

Figure 3- 24.Example of Pilkington Glass Manufacturing Data

3.7 Shading Devices

DOE-2 created the value of the profile angle to establish the position and the depth of the area upon which the solar rays will fall after passing through the envelope. The values of the profile angle are azimuth, altitude, and angle of Incidence of the Sun. Bearing of the Sun (Azimuth) is the direction of the sun with respect to Due South. The True Altitude of the Sun is the measure of the angle of the Sun from the building ground. The Angle of Incidence of the Sun with respect to a window or wall is need in estimating the solar heat gain.

There will be times when such a parametric shade is not suitable for a particular application. To allow for this, sun paths can be projected onto any surface from any point. This creates a **solar profile** line, which is clipped to the extent of the selected surfaces.

Once the position of the sun and incidence angles has been determined, the **solar exposure** of any surface can be calculated. This is done as either instantaneous irradiance or integrated irradiation over an entire day, month or year. In order to determine the amount of incident radiation, the following parameters are required: Global, beam and diffuse solar irradiance.

- Percentage shading or overshadowing
- Angular dependant reflectivity, if applicable
- Global, Beam and Diffuse Irradiance

Values for global and diffuse solar irradiance on a horizontal surface are calculated in two alternate ways depending on their use. When used as part of the thermal analysis of the model, hourly values for each day of the year are read directly from the location data file. These values are based on recorded meteorological weather data and are linked directly to cloud cover, sol-air temperature and absolute humidity.

When explicitly calculating instantaneous solar exposure for comparative analysis, recorded irradiation data can be misleading as values are used in isolation, with no information as to other external conditions such as cloud cover. In order to average out spurious fluctuations in these cases, values for global and diffuse irradiation are derived from average daily irradiation figures, as displayed in the monthly climate summary. The method used to derive instantaneous global and diffuse values this way is given by Szokolay. (Szokolay, S.V., 1987)

Once these two values have been read from the location file or derived from averaged monthly data, the amount of beam irradiance on a surface normal to the sun can be calculated using a method also described by Szokolay. To accurately quantity solar exposure, the effects of direct, diffuse and reflected irradiance must be considered.

The direct component is that irradiance which results from direct exposure to the sun. This depends on the angle of incidence of the **direct irradiance** as well as the percentage of the surface currently in shade.

Diffuse irradiance refers to that component of the total that arrives from all angles over the entire sky dome. This is dependant on the tilt angle of the surface. Obviously horizontal surfaces are exposed to the entire sky dome whilst vertical surfaces are exposed to only half. This is a simple linear relationship with tilt angle. The effects of geometric obstruction and overshadowing on the diffuse component are difficult to determine geometrically. An accurate method would require determining the average percentage of the sky dome visible from all points over the entire surface. The calculation of this value even for a single point is quite computationally intensive.

As a result it is assumed in this application that, whilst the diffuse component will have some effect on solar collection and sol-air temperature, this effect is not deterministically significant. Thus no consideration of overshadowing is applied to the diffuse component. This can lead to some anomalies within the model. For example, a floor element whose surface normal faces upwards would receive full diffuse radiation from the entire sky dome, even though almost entirely obscured. As the small amount of diffuse radiation that may impact on the floor is picked up by the windows that it must first pass through, objects defined as a FLOOR are not considered in diffuse radiation calculations. Similarly, any portion of a planar object that is adjacent to another zone is also not considered, regardless of the orientation of its surface normal.

Ground reflected irradiance is simply that component of the total that is reflected off the ground plane. This is dependant on tilt angle in the opposite way to diffuse sky irradiance. In the application, no consideration for geometric obstruction is applied to this component either. A default ground reflectivity of 0.2 is assumed unless explicitly set in the Preferences dialog box.

Solar exposure values can be graphed for any time of the day at any date. This information can be displayed as either hourly values throughout the day or as an average hourly distribution for each month of the year. These graphs show hourly values for total available irradiation, actual incident irradiation, any portion reflected off tagged surfaces and the percentage in shade for the selected object. Average hourly distributions can be shown for each of the above to understand when peak direct solar gains occur during the entire year.

Figure 3-25 shows position and proportion of overhang and fin relative to the window. In the base-case we assume from no overhang and fin, then starting to create to the ratio of 0.25, 0.50, and 0.75. For overhang or horizontal shade, the ratio is determined by the depth of overhang extending from the window wall, while the width is constant, divided by the height of the window. For Fin or Vertical shade, the ratio is determined by the depth of fin extending from the window wall, while the height is constant, divided by the height of the window.

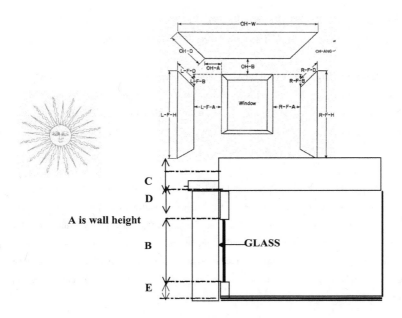

Figure 3- 25. Position and Proportion of Overhang and Fin Relative to the Window

Module has the shading devices such as over-hang and fin. Given a window and a limiting set of dates and times, it is possible to determine the exact geometry of an optimal shading device. DOE-2.1E has the sun path features that are extremely useful tool for both the site of a building and the design of shading devices. In the one diagram, overshadowing for an entire year can be displayed. These are constructed by projecting solar geometry the hemispherical sky dome onto a flat diagram, usually circular. A number of methods are available for translating solar altitude into a diagrammatic radius. The path of the sun through the sky dome is based on its actual altitude and azimuth calculated at each hour for the selected location Stereographic projection is the most widely used method and has been adopted in this application as it provides the greatest accuracy at low sun angles. A comprehensive analysis of alternate methods can be found in the Environmental Science Handbook. (Szokolay, S.V., 1980)

The method can be used to generate shades at any angle and with both horizontal and vertical elements. Optimized solar pergolas are also possible as the altitude of the sun when normal to the window can be determined at both the current date (used for the cut-off angle) and in mid-winter (used for maximum penetration). To model a building by the computer simulation methods, specific module's parameters and data have to be formulated and supplied to the simulation program. As a result the difference in the simulation output can be interpreted as having only been caused by the change in the input parameters. Table 3-7 shows the shading device size ratio.

Table 3- 7 Shading Device Size Ratio

Overhang Fin Ratio	Overhang Depth= Ratio * Window Height							
	Fin Depth= Ratio * Window Height							
	WWR= 10% Height=1.4	WWR= 30% Height=4.2	WWR= 40% Height=5.6	WWR =50% Height=7.0	WWR =60% Height=8.4	WWR =70% Height=9.8	WWR =75% Height= 10.50	WWR =80% Height= 11.20
0.00	0	0	0	0	0	0	0	0
0.25	0.35	1.05	1.40	1.75	2.10	2.45	2.63	2.80
0.50	0.70	2.10	2.80	3.50	4.20	4.90	5.25	5.60
0.75	1.05	3.15	4.20	5.25	6.30	7.35	7.88	8.40

Overhang Width =Window Width= 14 feet
Fin Height =Window Height

In order to develop simplified energy analysis and energy end use prediction, the investigations of design performance parameters range that lead to sophisticated modeling and simulation tools as part of new approach in the design process are necessary. The main focus is the development of these methods in the design decision-making. The architect is concerned with the following issues:

60

3.8 Space, Zone, and Surface

Continuous response to the space load and space study will create a customized building operation profile. The space occupancy pattern and schedule are extremely dynamic in changing due to the building operation: - they have a tremendous effect on building energy consumption. This scenario represents a load demand operation profile, which could be changeable even after the design is completed and the building is occupied

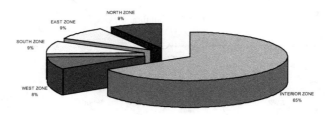

OFFICE BUILDING SPACE AND ZONE BY CATEGORIES

Figure 3- 26.Space and Zone by Categories

Figure 3-26 shows space and zone by categories. The surface area of a module is automatically calculated each time its geometry change as a result of rotation or the movement of its nodes. The algorithm here is to calculate and need to be able to handle module of any degree of complexity.

The module's floor area and the façade surface ratio can be simulated into computer worksheets and perform calculations. Charts and graphs produced from computer analysis of thermal gain or loss and lighting, enable the designer to determine according to north, south, east and west exposures.

Exterior surfaces of the module radiate thermal energy in all directions at all times. Interior surfaces consists of partition, exchange the heat between its adjacent space and surfaces. Thermal radiation consists of long-wave radiation emitted at low temperature, whereas solar radiation is short-wave radiation emitted at very high temperatures.

3.9 Infiltration Rate

Infiltration Rate can be either Air Change Per Hour or CFM/SQFT. The rate used was office infiltration at 0.06CFM per linear Foot of window perimeter, which is equal to 2 * (15+window height) * 0.06 = X CFM (varied according to window to wall ratio). Infiltration of cfm/sq.ft is X CFM/module area = X/225. The Air Change Per Hour is X * 60minutes per hour/module volume= X * 60/3150. If percent of window glass opening equals to 10%, air change per hour is 0.052, if 40% air change per hour is 0.066, if 50% air change per hour is 0.070, if 60% air change per hour is 0.075, if 80% air change per hour is 0.084. Infiltration schedule is different according to the climate location. For example Chicago has 100% air infiltrate during the wintertime, 50% during the spring and fall, while in summertime there is no infiltration through the building.

3.10 Operating Schedule

Operating schedule factors are determined from the heat gain from occupants, lighting, infiltration, and heat generating equipments, according to the field actual space study for the existing office buildings both in Chicago and Bangkok. The space occupancy pattern and schedule are extremely dynamic and changing due to the building operation, which cause tremendous effect to building energy consumption. Working people, staffs and visitors are primary users of the building. They are considered non-weather dependent loads, or base energy profile. In this study used only lighting schedule to see the effect of day-lighting and building envelope. From the base-case module, we can make an assumption that the building is opened for the entire year except official holidays; regular time schedule is from 8am-6pm Monday-Friday and 8am-6pm on Saturday. This scenario represents load representative demand operation profile, which could be changeable even after the design is completed and the building is occupied. The lighting schedule used is office schedule, opened for the entire year except official holidays; regular time schedule is from 8am-7pm Monday-Friday and 8am-6pm on Saturday. Figure 3-27.shows example of building operating hours schedule. Table 3-8 contains example of lighting schedule used in the module.

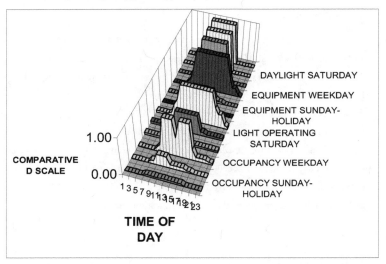

Figure 3- 27.Example of Building Operating Hours Schedule

Table 3- 8 Example of Lighting Schedule Used In the Module

AM PM	1	2	3	4	5	6	7	8	9	10	11	12
Week Day	0.05	0.05	0.05	0.05	0.05	0.10	0.35	0.50	0.90	0.90	0.90	0.90
Week Day	0.90	0.90	0.90	0.90	0.90	0.90	0.90	0.50	0.35	0.35	0.10	0.05
Saturday	0.05	0.05	0.05	0.05	0.05	0.05	0.10	0.10	0.50	0.50	0.50	0.50
Saturday	0.50	0.50	0.50	0.50	0.10	0.10	0.05	0.05	0.05	0.05	0.05	0.05
Sunday	0.05	0.05	0.05	0.05	0.05	0.05	0.05	0.05	0.05	0.05	0.05	0.05
Sunday	0.05	0.05	0.05	0.05	0.05	0.05	0.05	0.05	0.05	0.05	0.05	0.05
Holiday	0.05	0.05	0.05	0.05	0.05	0.05	0.05	0.05	0.05	0.05	0.05	0.05
Holiday	0.05	0.05	0.05	0.05	0.05	0.05	0.05	0.05	0.05	0.05	0.05	0.05

3.11 Building Construction and Materials

All envelope materials radiate thermal energy in all directions at all times. This thermal radiation consists of **long-wave radiation** emitted at low temperature, whereas solar radiation is **short-wave radiation** emitted at very high temperatures. Envelope materials' ability to absorb thermal radiation depends on its **surface density, the component, and incident angle of thermal radiation** being parallel to the material surface. The measure of the envelope materials' ability to give off radiant heat is referred to as the **material emissivity (E).**

Typical wall construction consists of heavy weight concrete and concrete blocks with cement plaster finished. Floor can be the 4-inch concrete floors and tiles. The module contains 225sq.ft.of floor area. Ceilings are some ¾ inch applied acoustic tiles and some are suspended acoustic tile. Partitions are gypsum boards that stimulate convective heat transfer between core and perimeter spaces using a U-Value of 0.50 Btu/hr.-sq.ft-°F. Figure 3-28 shows the temperature gradient in wall materials. Figure 3-

29 shows periodic heat flow of brick wall and a Zero Mass Element, while a mass wall yields time lag and better balance point. Table 3-9 contains heat transfer and emissivity values of building materials.

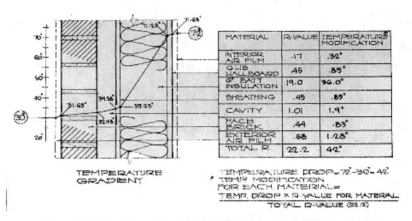

Figure 3- 28. Temperature Gradient in Wall Materials (Olygay, "Design With Climate")

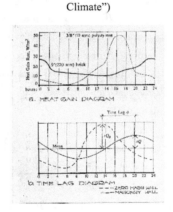

Figure 3- 29.Periodic Heat flow of Brick wall and a Zero Mass Element, Mass Wall Yields Time Lag and Better Balance Point.

Table 3- 9 Heat Transfer and Emissivity Values of Building Materials

Surfaces	Normal Emissivity Thermal Radiation	Absorption Solar Radiation
1. Brick	0.90-0.98	0.85-0.98
2. Concrete	0.85-0.95	0.65-0.80
3. Stone	0.85-0.95	0.50-0.70
4. White Brick Tile	0.85-0.95	0.30-0.50
5. Glass	0.90	0.04-0.40
6. Polished Metallic	0.02-0.04	0.10-0.40
7. Dull Metallic	0.20-0.30	0.40-0.65

3.12 Natural Day-lighting and Artificial Lighting

In the base-case we assume from no daylight and then starting to bring daylight in according to the perturbation and daylight schedule. Natural day-lighting is important issues by the way they were introduced into the buildings, which is a combination of direct, reflected or refracted and diffuse light introduced into buildings in the following ways.

- Control governed by the available day lighting to a set point of 50-foot candles.
- Assumed daylight hours are 6am-6pm all year.
- Allow Day lighting to penetrate into the module
- Day Light-Set Point (to offset artificial lights) = 50 foot-candles
- Light Reference Point = 7.5feet, 5feet, 3feet
- Daylight-Schedule 6am-6pm

Artificial lighting levels are calculated using the point-by-point method. This is a geometric algorithm in which the contribution of each light source is simple summed at points within the enclosure. The luminance distribution of each light source is used to modify contributed levels based on the off-axis angle of each point. The inverse square law is then used to account for geometric spreading. This is a simple method for determining direct light levels from both regularly and irregularly spaced luminaries. It is a useful tool for ensuring adequate minimum light levels and is mainly intended for use in this application as a comparative measure. This method takes no account of diffuse light or inter-reflection between illuminated surfaces.

The point-by-point method can be used in place of the lumen method to design and analyses regular arrays as well as localized lighting. Lighting in the module used the followings:

- Lighting W/SQFT = 2 Watts/ft.²
- Light-To-Space = 1.0
- Lighting-Type=Fluorescent

3.13 Variation of Placing the Module in The Buildings

As module change according to the design parameters, the energy results change. In order to develop simplified energy analysis for the building, the investigations of design performance parameters range that lead to sophisticated modeling and simulation tools are necessary.

This research will show the case study of the prototype theoretical buildings as the way to present method and process of how to do, The study evaluates environmental performance that exists among the number of most possible prototype module can be placed in different configurations. A wide range of values was studied for each parameter even this range might not reflect the current possibilities that found in the actual buildings. Figure 3-30 shows variation of placing the module in the building.

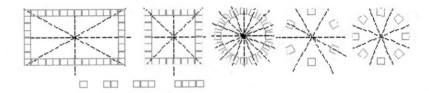

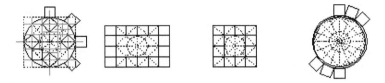

Figure 3- 30. Variation of placing the Module in Buildings

CHAPTER 4

BUILDING ENERGY EFFICIENT MODELING TOOL-MODELLER

A simplified energy tool for architects is developed from utilizes a database of DOE2.1E simulations imposed on the module (Chapter 3) in order to predict the building energy performance. The DOE2.1E building energy analysis program performs a detailed hourly simulation of energy use in buildings. It provides a mechanism for entering and editing building information in the Building Description Language (BDL), the input format for the DOE2 simulation engine. DOE2 development was initiated as a collaborative effort between the U.S. Department of Energy's Office of Building Technologies at Lawrence Berkeley Laboratory (LBL), and the Electric Power Research Institute (EPRI).

The layout of the interface for the tool called "Building Energy Modeler" is intend to be simple enough so architects would not have to spend much time to learn or need to need to consult the document in order to use, and can concentrate on their architectural works. Help files are nevertheless provided and accessible through information screen located right below the input pull down menu. The interface, which confronts architects contain the graphic screen with many controls and buttons. Architects must specify the input parameters that most related to their design in order to created the module that represents the mass or shape of the proposed building. In the mode of energy performance selections, architects choose the one(s) they want to view the building energy performance that was automatically pre-calculated by DOE-2 and set to be consistent with the chosen architectural input parameters or elements. Architects can change the climate or building location and study the performance of the façade in any orientation. The software feature provides the illustration of the program operation and a benchmark, which architects can compare with their own building.

4.1 Define the Tool

The term design tool is generally applied to a wide range of techniques, from the use of tabulated data sheets and manual calculation methods, through to sophisticated computer analysis software. In the context of this work, the term is used to describe computer software developed to replace laborious manual calculations used to inform the design decision-making process. Using a computer to perform the mathematical component makes it possible to study effects not previously considered in many building design.

While the use of design tools may not automatically eradicate high safety margins and over-design, experience shows that they can significantly increase a designer's confidence in the solution. Most design margins result from the perceived application of a generalized solution to a very specific problem. As the use of most advanced simulation tools begins with the entry of a very detailed model, the results produced at least appear to be more specific to the actual problem they are being applied to.

Modeler software is divided into 7 main sub modules: heating, cooling, electrical lighting, peak loads, load components, and day lighting. The heating and cooling sub modules stimulated the exchange of heat gain and loss through the skin envelope of the building, and show the annual required energy of heating and cooling. Day lighting sub module predicts the distribution of sunlight penetrates through the window of the module, according to the time of year and location, and calculates yearlong average daylight projected into the work (sensor) surface in the room. Electrical lighting sub module makes up for the amount of lacking lighting energy from what natural daylight can offer. Since he software retains the weather data for every hour of the year, architects may interface the sub modules in any cities where their projects situated on hourly basis.

The followings are the characteristics and features of this tool
- Basis for Evaluation and Comparison
- Model and analysis for thermal performance.

- Tool for conceptual stage of design. User can change various attributes or parameters.
- A parametric analysis is undertaken to delineate sensitivities
- Lighting Analysis and Simulation
- Shadow analysis that incorporated into the results from overhangs and fin parameters.

Principle of the energy efficient design is based on the design parameters performance and interaction. The architect will successfully deal with climate and environment in a better way by comparison between the two schemes or more and will be able to evaluate according to the outdoor climatic condition with indoor comfort.

4.2 Thermal Performance Analysis Model

Heat flow calculation method in engineering was already applied to the designed Modeler software, and determines heating cooling and lighting loads. DOE-2.1E used in this research is FORTRAN based programming and uses BDL language as an input and command keys. The result or outcomes from DOE-2.1E are statistically valid and widely used in research and professional practice.

Through the parametric study process, it is possible to evaluate the effects of the architect's envelope. A number of results including reports, table, graphs, charts or graphics formed from the simulation process can be provided to aid the architect's decision making. For all energy-generated factors, the simulation results of energy consumption indices can provide architects with references for the energy demand of the buildings they design and enable them to set targets for the energy and environmental performance the project should have.

4.3 Conceptual Design Model

The conceptual stage of design occurs very early in the design process. This is the time when a vast array of competing requirements shapes the initial building form, when geometry, materials and orientation are still being formulated. The use of module can overcome the detailed nature and amount of input required to describe the building model and simulate its performance. Having to enter this data very early in the design would interrupt the process of iterative decision-making and prematurely force the designer to a series of arbitrary decisions. The conceptual design tool must make the process of entering the data part of the design process. The key is to produce real and useful design feedback at every stage of the modeling process from data entry right through to final analysis. These requirements are placed firmly on the interface that is the means by which the user describes and interacts with the model.

4.4 Parametric Analysis Delineates Sensitivities

In this dissertation, a computational model is developed to study the heat transfer behavior and energy consumed by building components, of any given projects that designers have to face. A parametric analysis is undertaken to delineate sensitivities in order to control and measure parameters such as climatic condition, building orientation, percentage of glass to surface ratio, glazing types, exterior shading devices. Predicting the changes in energy usages and energy performance as a result of changes in parameters is critical in determining the way to design the building schemes, which will lead to better understanding of subsequent energy control strategies.

4.5 Lighting Analysis and Simulation

More sophisticated methods of lighting analysis make use of full rendering techniques. The aim when using more advanced lighting simulation tools is to produce realistic images of the spaces within a geometric building model that correspond as closely as possible to what would actually be found if that space were real. There is a major difference between photo-realistic rendering and actual lighting simulation, use physically realistic values for the light sources or the surface reflectance, to derive more complex information such as luminance levels, daylight factors, glare indexes and even visual comfort.

Daylight factor can be calculated manually using the British Research Station (BRS) Daylight Factor Protractors or lighting applications such as Adeline from the International Energy Agency (IEA). Daylight Factor is the ratio of the daylight illumination at a point on a given plane due to light received directly or indirectly from a sky of assumed or known luminance distribution, to the illumination on a horizontal plane due to an unobstructed hemisphere of this sky. Direct sunlight is excluded for both values of illumination. (Longmore, 1968)

While the accurate calculation of such levels is quite computationally intensive, a manual method based on the British Research Station (BRS) Daylight Factor Protractors is widely used as a first approximation. The Daylight Factor at a point within an enclosure is a function of three components, the sky component, externally reflected component and internally reflected component. (Longmore, 1968) The calculation of Daylight Factor in this application uses a geometric method for determining the sky and externally reflected components and the standard BRS formulae for the internally reflected component

The aim of lighting simulation is to be able to predict lighting levels at points within a module. The process of accurately calculating these levels can become quite involved given the complex nature of surface inter-reflection. Calculating natural light levels is more difficult as diffuse light from the sky dome itself is often the main source, not just the direct sun. As cloud cover is a significant variable, DOE-2.1E developed method of daylight estimation, and a model of the distribution of light over the sky, for overcast, uniform and sunny conditions. An estimate of day lighting levels and glare within a building is fundamental to fenestration design.

If the daylight at a particular sensor point and the design sky luminance for the current location is known, then a representative value for illumination for a uniform or overcast sky can be calculated. In our case the sensor was placed 7.5 feet from left hand side from the width of the window (15 feet), 5 feet deep from the perimeter window wall, and 3 feet above the ground. The resulting value is neither a minimum nor an average, but a representative design value of useful interior daylight available over approximately 85% of the working year. This relationship is useful as it can be reversed to actually determine a daylight factor. If the required illumination for a particular task is known, then the daylight factor required providing that level 85% of the time is simply given by:

Daylight Factor = Design Sky Illuminance / Required Illumination

4.6 Shadow Analysis

An adjunct to lighting simulation is the analysis of sun penetration and solar shading for determination of the size of overhang and fin. DOE2.1E perform the parametric energy analysis, based on the analysis and equations take into account parameters that change in their own group, and the interactions that occur when parameters are changed simultaneously.

The software tool was developed using a created database generated from DOE-2.1E results; it also provides parameter evaluation, as a method of investigating alternatives, and direct information about the relationship between different energy components. It also offers designers the prescriptive simplified graphs and charts that interpreted from the quantitative simulation achievable in different energy components as they relate to the building and to each other.

4.7 Database Queries and Simulation

The components that will be used in the database include:

- Forms to accept the data and to assist with database navigation
- Reports that can be used to print out information
- Queries and macros that help analyze the data for the architects to use, and prepare for export into energy design software programs.
- Capability to get the updated, in-depth information architects need from the simulation database.
- Capability to search and retrieve information.
- Capability to link with databases and lookup data

4.8 The Software Program Components

The building and built environment design process is an iterative problem solving process. It is difficult for the designer to solve all aspects of environmental problem since it is always possible to generate a new solution. Thus limits the ability of architect to solve complex problems that involve many environmental variables. This program is more design guideline tool than analysis, because it generates alternatives that help to process energy to sustainability information more productivity. This program can begin when the building is only vaguely defined, produce the energy use results in seconds in order to maintain the momentum of interactive process, ability to generate graphics from simulation or numerical calculation results, and capable of indication the energy for sustainable solutions between what is better or worse than another (2 schemes). The ideal of this program is easy to use and precise or accurate information. People will become familiar very rapidly and there is instruction built-in help menu to use. If they need support or have questions, there is an email attached to that they can write to seek help.

Figure 4-1 shows the characteristics of this software tool. Figure 4-2 shows software design process. The followings summarize the characteristics of the program:

- Easy to use, easy to learn.
- The learning period is approximately half an hour.
- The documentation required is the built-in help in the program.
- Has an important graphical component of the required data input to produce useful energy results.
- The users need not to be expert in this field.
- Sufficient precision.
- Accessible to large number of users.
- Can become available in public domain through Internet in order to reach large number of people.
- Can be downloaded to the personal computer or use in the web browser (for example the URL of sustainability focusing on energy efficiency).
- Simplest interface

Modeler software

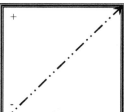

Easy to use   *precise tool*

Figure 4- 1Software tool

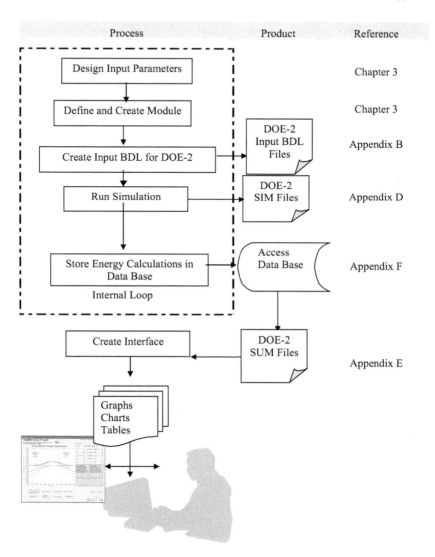

Figure 4- 2 Software Design Process

78

The software is an energy design interface tool for architects, providing very detailed calculation results from DOE-2 and presenting in graphical formats that simplify for architects to use during the schematic design process, and simulating the method to help architects find the comparative solutions. Figure 4-3 shows the software components

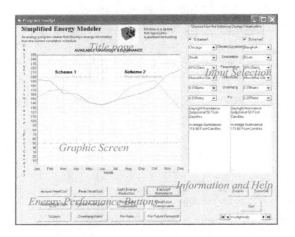

Figure 4- 3 Software Components

4.8.1 Software Title. Software Title informs the current version of the program.

4.8.2 Design Parameter Selection Toolbars. Architects can choose the design parameters that they are interested to investigate from the available options; climate location, orientation, percent of glass, glass types, overhang and fin shading devices, and be able to compare the 2 schemes. The toolbar selection is intended to use to correlate the results of energy performance with the changing design parameters. Combination is the result of the parametric interaction level; from first level (base case) up to sixth multiple parameter combination level.

4.8.3 Graphic Screen. Graphics screen shows the results in graphs and charts, the user can move the mouse over to see the selected result in exact number.

4.8.4 Information Screen and Help Files. Information screen and help files show the data and information in numerical, graphic, and text formats. Note: Help files give instruction how to use the program and interpret the meaning of the results

4.8.5 Building Energy Performance Buttons. There are 11 command buttons that architects can select to view the energy performance based on their interest. This interface covers all the heat gain, heat loss, daylight, and light loads that architects need to work on their design schemes.

4.8.6 Utility Tools. Utility tools consist of zoom in and out tools, and exit the program tool.

4.9 Building Energy Performance

The following results are based on real simulation of two strategies for Chicago; scheme1 used envelope strategy, which represents south facing orientation, 50% lowE clear glass, without overhang and fin, scheme2 used shading strategy, which represents south facing orientation, 80% monolithic clear glass, 0.50 ratios of both overhang and fin.

4.9.1 Peak Heating, Cooling, and Lighting Loads. Peak Thermal Loads is the energy that flows through the skin or envelope at peak condition, results of this is Btu/hr or percent. Once a module is placed, internal temperatures and load for the module can be determined from actual recorded weather data. Standard practice is by summing the peak loads in each time the module is placed and applying a simple diversity factor to account for sun movement. Using more sophisticated tools, peak heating and cooling loads can be determined from maximum hourly loads in the module, taken over the entire year. Such loads take full consideration of fresh air requirements based on actual outside air conditions as well as changes in level and usage throughout each day. This level of accuracy, and the potential for increased efficiency at each component, can result in significant energy savings. The followings are the output results that can obtain from simulation. Figure 4-4 shows the graphs. Examples can be seen in the Appendix.

- Annual Peak Cooling Load (KBtu/Hr); Time of peak; Out door Dry-bulb and Wet-bulb temperature; Indoor temperature is 72°F
- Annual Peak Heating Load (Kbtu/hr); Time of peak; Out door Dry-bulb and Wet-bulb temperature; Indoor temperature is 72°F
- Total peak load; index is total peak load / area = total peak load / 225 (Kbtu-hr/sq.ft); multiply the index by total net floor area to obtain total building peak load ((Kbtu-hr)
- Note: Total annual energy demand (KW) = Total building peak load * 0.2931

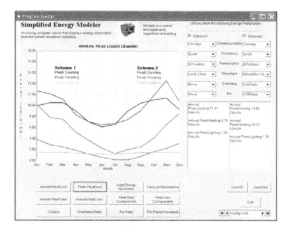

Figure 4- 4 Peak Heating, Cooling, and Lighting Loads

4.9.2 Load Components. Thermal loads for all of the important energy generating components are wall, glass conduction, glass solar radiation, infiltration, and lights. Peak load for each component is either cooling load or heat gain is situation where the building is over-heated; or heating load or heat loss is situation where the building is under-heated. Glass conduction is the sum of the U*A*ΔT heat gain through all the windows in the space plus solar energy absorbed by the glass and conducted into the space. Glass solar radiation is the heat gain caused by direct and diffuse solar radiation transmitted by the window into the space. Figure 4-5 and 4-6 shows the graphs.

Note: all sensible loads are calculated as delayed in time with weighting factors, and it is possible to have heat gain from glass solar radiation, for example long after the sun has moved and no longer shines on the exposed walls of the space.

82

The sensible portion of infiltration is treated as instantaneous heat gain or loss. The latent

portion is reported and passed into the system as a CFM with the calculation of humidity

ratio for each hour. The contribution of the latent heat (negative or positive in relation of

the room humidity) is then calculated as a mass balance of moisture in the space, to

determine the return air humidity ratio. Chicago is dryer climate than Bangkok; the

infiltration may result in a decreased space latent load and total system load, where in

Bangkok infiltration acts to increase system load.

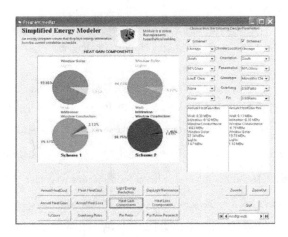

Figure 4- 5 Summer. Heating and Cooling Load Components

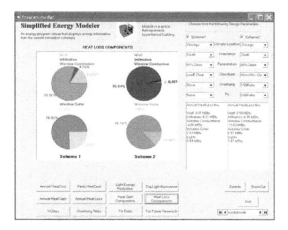

Figure 4- 6 Winter heating and Cooling Load Components

4.9.3 Annual Heating and Cooling Loads Requirements. Cooling, heating, and lighting are the main three parts that based on constant temperature. Annual Thermal Loads: are the annual thermal loads that the building mechanical and environmental system will have to handle during the internal cooling and heating period. Results (Btu or percent) are a true indication of annual cooling and heating loads for the skin or envelope thermal gains.

The followings are the output results that can obtain from simulation. Figure 4-7 shows the graphs. Examples can be seen in the Appendix.

- Annual cooling Requirement (Mbtu)
- Minimum Monthly Cooling Load (Mbtu)
- Maximum Monthly Cooling Load (Mbtu)
- Annual Heating Requirement (Mbtu)

- Minimum Monthly Heating Load (Mbtu)
- Maximum Monthly Heating Load (Mbtu)
- Energy savings for cooling (MBtu), Scheme1-Scheme2 (in case Scheme2 is better)
- Energy savings for heating (Mbtu), Scheme1-Scheme2 (in case Scheme2 is better)
- Electrical Demand Consumption
- Note: Balance point is when the zero energy; occurred when Cooling=Heating.

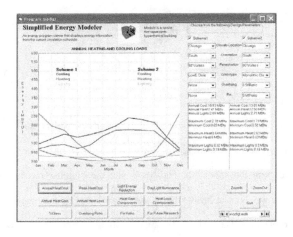

Figure 4- 7 Monthly Heating and Cooling Loads

4.9.4 Monthly Heat Gain and Heat Loss Components. Ratio between heat gain and heat loss equals to Total heat gain divided by Total heat loss. Monthly Heat gain through each components equals to monthly load for each components (Mbtu), which consists wall, window conduction, window solar radiation, infiltration, and artificial Lights; cooling load or heat gain is situation where the building is over-heated; heating load or heat loss is situation where the building is under-heated. Monthly Heat loss through each components equals to monthly load for each components (Mbtu), which consists of wall, window conduction, window solar radiation, infiltration, and artificial lights; cooling load or heat gain is situation where the building is over-heated; heating load or heat loss is situation where the building is under-heated. Figure 4-8 and 4-9 shows the graphs

The followings are the output results that can obtain from simulation:

- Components that contributed the highest heat gain.
- Components that contributed the lowest heat gain.
- Component that contributed the highest heat loss
- Components that contributed the lowest heat loss.
- Scheme (1 or 2) that yields better or worse heat gain in the summer.
- Scheme (1 or 2) that yields better or worse heat loss in the winter.
- Annual heat gain contributed for each component.
- Annual Heat loss contributed for each component.

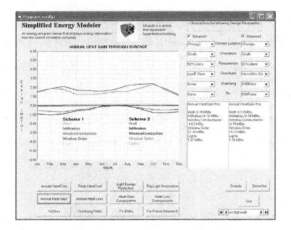

Figure 4- 8 Monthly Heat Gain Components

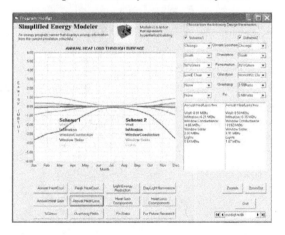

Figure 4- 9 Monthly Heat Loss Components

4.9.5 Percent Lighting Energy Savings by Daylight. Day lighting and energy savings

Index's results are in percentage and present in annual energy savings, since savings can effect operating cost and economic performance over life time. Potential amount from daylight will be able to substitute electrical lighting use based on the hourly profile (daytime versus nighttime). This feature allows determining what affect the use of day lighting to dim electrical lighting has on energy use, peak loads, and energy cost. Figure 4-10 shows the graphs. The followings are the performance that can obtain from simulation and more:

- Maximum lighting energy savings by daylight (%)
- Minimum lighting energy savings by daylight (%)
- Scheme (1 or 2) that yields better lighting energy savings.
- The followings are the output results that can obtain from simulation:
- Month(s) that has maximum lighting energy savings by daylight.
- Month(s) that has minimum lighting energy savings by daylight.

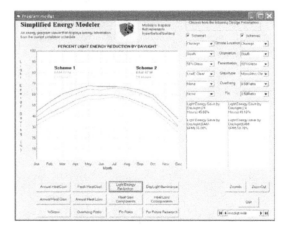

Figure 4- 10 Percent Lighting Energy Savings by Daylight

4.9.6 Daylight I luminance and Set Point. Luminance contribution and distribution are found by interpolating the stored daylight factors using current hours sun position and cloud cover, then multiply by current hour exterior horizontal luminance obtained from measured horizontal solar radiation. Sensor was placed at X position (middle of the window wall) equals to 7.5 ft.; Y position (deep into the space from perimeter wall) equals to 5.0 ft; Z position (height above from the floor) equals to 3.0 ft. Light control type is continuous; light set point is 50 foot-candle. Figure 4-11 shows the graphs. The followings are the output results that can obtain from simulation:

- Illumination level
- Illumination range
- Daylight Factor
- Schemes (1 or 2) that yields higher daylight luminance.
- Maximum daylight luminance occurred inside the module.
- Minimum daylight luminance occurred inside the module.
- Number of months is above set point (Set point equals to 50 foot candles).
- Number of months is below set point (Set point equals to 50 foot candles).
- Monthly Day-lighting I luminance Frequency of Occurring

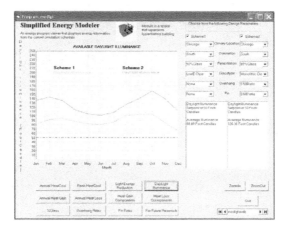

Figure 4- 11 Daylight Illumination

Note: In order to get the explanation of the graphs, charts, and data in table format, look at the advisor menu in the software program.

4.9.7 Percent of Glass. This feature provides the performance of solar heat gain and lighting usage for 10% to 80% glass. DOE-2 integrates the area of each window size to obtain the contribution of direct sunlight and solar radiation and the diffuse sunlight from sky and ground. Taken into account are such factors as window size, orientation, glass transmittance, position of the sun, and cloudiness of the sky. Figure 4-12 shows the graphs of percent glass.

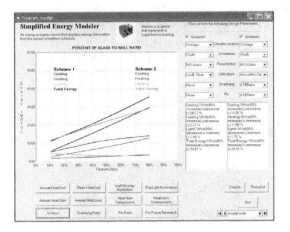

Figure 4- 12 Percent Glass

4.9.8 Overhang Ratio. This feature provides the performance of solar heat gain and lighting usage for different size ratio of overhang. If combine overhang and fin together would get an egg crate. Figure 4-13 shows the graphs of overhang ratio.

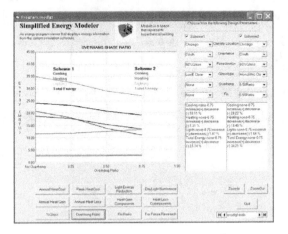

Figure 4- 13 Overhang Ratio

4.9.9 Fin Ratio. This feature provides the performance of solar heat gain and lighting usage for different size ratio of fin. If combine overhang and fin together would get an egg crate. Figure 4-14 shows the graphs of fin ration.

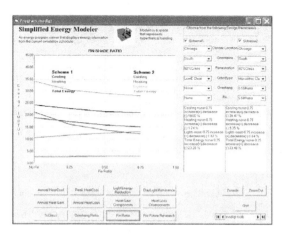

Figure 4- 14 Fin Ratio

CHAPTER 5

CASE SCENARIOS: GUIDELINES HOW TO USE THE SOFTWARE

This software tool helps architects design and plan while utilize natural possibilities in a given climate setting (Chicago and Bangkok), depend on how much the architectural features or elements can alter the actual thermal situation. In summer time usually the temperature outside is warmer. Heat captured with lessen heat gain kept in over-heated periods will reduce cooling budget. In winter time usually the temperature outside is colder. Heat captured with lessen heat loss in under-heated periods will reduce heating budget. If heat gains balance with heat loss then architects can achieve zero energy use for their buildings. Architects can either use or increase shading to reduce solar gain in the summer or changing percent of glass or glass types to achieve the same objective, or changing orientation to accumulate heat in winter.

It remains necessary for architects to correlate the findings. The findings must be brought together in the way, which architectural elements work together. It also makes it possible to evaluate situations, which overlap or run against each other, indicating measures should be taken in a specific situation. The use of efficient tool and technique can help in analyze complex design problems. The followings targets on using this software program solving the design approach responsive building and built environment sustainability through energy efficiency and savings strategies.

Choose the given climate location; Chicago or Bangkok. Climate data of these specific regions were already analyzed with the yearly characteristics of their constituent elements; temperature, relative humidity, solar radiation, and wind effects.

5.1 Case Scenario 1: Optimize the Site Selection and Orientation. Climate Location; Chicago

In the site selection and site planning, sites at which orientation had desirable and showed winter-summer relationship with minimum heat gain in summer and the lowest heat loss in winter. A balance can be found between under-heated period when we seek solar radiation, and over-heated period when we want to avoid with need of cooling loads. The climate balance curves represent the summation of reconstructed heat flow curves. Optimum orientation is defined as minimum as possible summer heat gain (or maximum as possible summer heat loss) during over-heated periods associated with as maximum as possible winter heat gain (or minimum as possible winter heat loss) during under-heated periods

Figure 5- 1 Chicago Climate, Orientation, Maximum 80% Monolithic Clear Glass

In orientation for the buildings the solar heat is both positive (heat gain in hot periods-shown in blue color in the graph), and negative (heat loss in cold periods-shown in red color in the graph). A balance can be found between under-heated period when we seek solar radiation, and over-heated period when we want to avoid.

94

Architects' task is to position a building so as to take the best advantage of the climate balance and sun's values for thermal effect, hygiene, and psychological benefits. Optimum orientation is defined as minimum as possible summer heat gain (or maximum as possible summer heat loss) during over-heated periods associated with as maximum as possible winter heat gain (or minimum as possible winter heat loss) during under-heated periods.

5.1.1 Problem definition. Architects design the new project in Chicago and plan to utilize 80% widow wall façade with the single clear glass. They want to know which orientation is the best for this project site. Architects' task is to position a building so as to take the best advantage of the bioclimatic balance and sun's values for thermal effect. Figure 5-1.shows the definition of the problems.

5.1.2 Strategies. Methodology is in both Peak Heating and Cooling and Annual Heating and Cooling loads. Find maximum heat gain or minimum heat loss during under-heated periods in winter, and minimum heat gain or maximum heat loss during over-heated period in summer. Analyze daily or monthly solar radiation characteristics and climate balance.

5.1.3 Steps to do

- In the software program for both schemes, under climate location pull down box choose Chicago, under fenestration choose 80% glass, under glass type choose monolithic clear glass, choose none for both overhang and fin.
- In the scheme 1 under orientation choose North as a base case for comparison (can be any orientation for a base case as you wish)
- In the scheme 2 under orientation choose Northeast

- Click Peak Heat Cool command button to obtain the peak solar radiation results in the graphic screen.
- Click the Annual Heat Cool command button to obtain the solar heat gain and loss results in the graphic screen.
- Compare scheme 1 and 2 and choose the best one.
- Continue with Northwest orientation
- Repeat step 4-6
- At this step, it can be known that should go clockwise
- Continue with East, Southeast, South, Southwest, West orientation
- Repeat step 4-

96

To start with the comparison between 2 schemes, it can start from North or from South, and compare to the next scheme Northeast and Northwest or Southeast and Southwest - and go to direction clockwise or counter clockwise depend on which orientation has better climate balance and desirable for the site. Figure 5-2 to 5-7 compare peak radiation and annual heat gain and heat loss for different orientation.

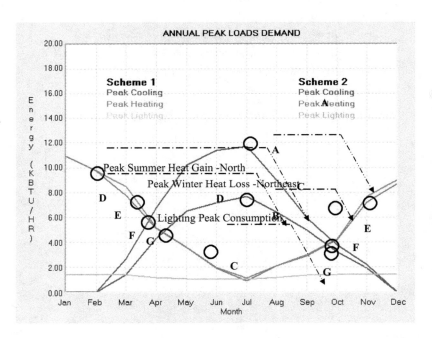

Figure 5- 2 Peak Radiation Heat Gain and Heat Loss-North-Northeast

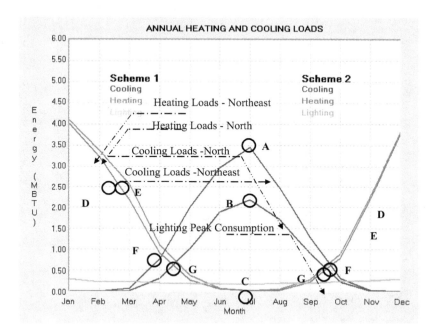

Figure 5- 3 Annual Radiation Heat Gain and Heat Loss-North-Northeast

98

Starting from North, continue from 1 to 6

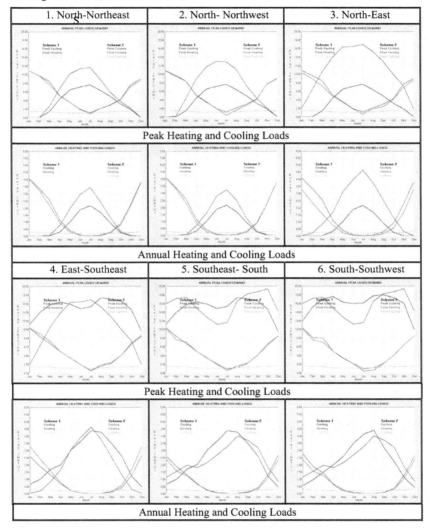

Figure 5- 4.Compare Peak and Annual Heat Gain and Heat Loss-Start from North

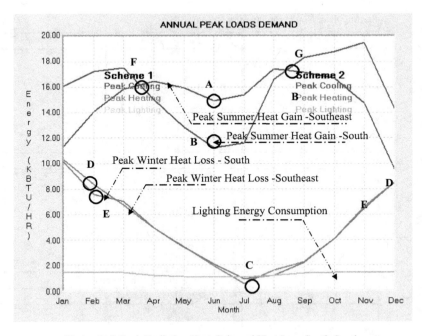

Figure 5- 5 Peak Radiation Heat Gain and Heat Loss South-Southeast

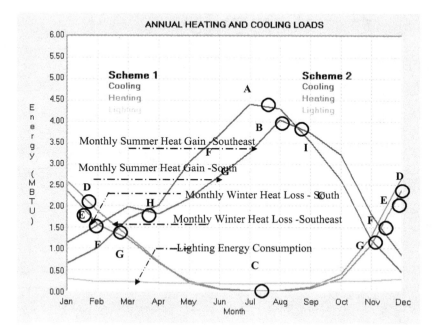

Figure 5- 6 Annual Radiation Heat Gain and Heat Loss South-Southeast

102

Starting from South, continue 1 to 6

| 1. South- Southeast | 2. South-Southwest | 3. South-West |

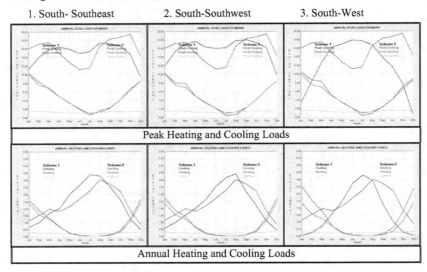

Peak Heating and Cooling Loads

Annual Heating and Cooling Loads

Figure 5- 7.Compare Peak and Annual Heat Gain and Heat Loss-Start from South

5.1.4 Analysis

- Solar radiation occurred during overheated periods and under heated periods. North has lower summer heat gain (B) than Northeast (A), while Northeast has lower winter heat loss (D) than North. (E). The difference between A and B is cooling loads saving. F and G are Balance Points. North also has better balance point (G) than Northeast (F).

- North has lower peak summer heat gain (B) than Northeast (A). Northeast has lower winter heat loss (D) than North. (E).The difference between A and B is peak cooling loads saving. F and G are Balance Points. North also has better balance point (G) than Northeast (F).

- South has lower peak summer heat gain (B) than Southeast (A) and has lower winter heat loss (E) than Southeast (D). The difference between A and B is peak cooling loads saving. F and G are the free cooling.

    - South has lower summer heat gain (B) than Southeast (A) and lower winter heat loss (E) than Southeast (D). The difference between A and B is cooling loads saving. South has better climate balance (F) and than Southeast (G) and free cooling (H) and (I).

5.1.5 Solutions. Figure 5-10 shows the summary of the results of optimize diagram.

- South is the most desirable.
- Minimize energy consumption and maximize energy savings.
- South has the minimum summer heat gain (cooling loads) and winter heat loss (heating loads).
- South has the best climate balance between summer heat gain (cooling loads) and winter heat loss (heating loads).

Less Desirable   More Desirable

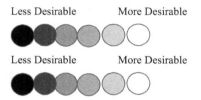

Less Desirable   More Desirable

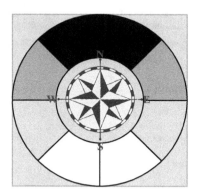

Figure 5- 8 Pattern of the Most Desirable Orientation for Site Selection

104

5.2 Case Scenario 2 Optimize the Site and Orientation with the Climate Balance

Climate Location: Bangkok

Bangkok is in tropical climate, so the main concerns, which different from Chicago are heat gain in over-heated periods and humidity. Optimum orientation is defined as minimum as possible summer heat gain (or maximum as possible summer heat loss) during over-heated periods associated with as maximum as possible winter heat gain (or minimum as possible winter heat loss) during under-heated periods.

Figure 5- 9 Bangkok Climate, Orientation, Maximum 80% Monolithic Clear Glass

5.2.1 Problem Definition. Architects design the project for another climate location in Bangkok, and plan to utilize 80% widow wall façade with the single clear glass. They want to know which orientation is the best for this project site. Figure 5-9 shows the definition of the problems.

5.2.2 Strategies. With the climate balance method, examine both Peak Heating and Cooling and Annual Heating and Cooling. Find maximum heat gain or minimum heat loss during under-heated periods in winter, and minimum heat gain or maximum heat loss during over-heated period in summer. Analyze daily or monthly solar radiation characteristics and climate balanced heat distribution.

5.2.3 Steps to do

- In the software program for both schemes, under climate location pull down box choose Bangkok, under fenestration choose 80% glass, under glass type choose monolithic clear glass, and choose none for both overhang and fin.
- In the scheme 1 under orientation choose North as a base case for comparison (can be any orientation for a base case as you wish).
- In the scheme 2 under orientation choose Northeast
- Click Peak Heat Cool command button to obtain the peak solar radiation results in the graphic screen.
- Click the Annual Heat Cool command button to obtain the solar heat gain and loss results in the graphic screen.
- Compare scheme 1 and 2 and choose the best one.
- Continue with Northwest orientation
- Repeat step 4--6
- At this step, it can be known that should go clockwise
- Continue with East, Southeast, South, Southwest, and West orientation.
- Repeat step 4-6

To start with the comparison between 2 schemes, it can start from North or from South, and compare to the next scheme Northeast and Northwest or Southeast and Southwest--- and go to direction clockwise or counter clockwise depend on which orientation has better climate balance and desirable for the site. Figure 5-10 to 5-15 compare peak radiation and annual heat gain and heat loss for different orientation.

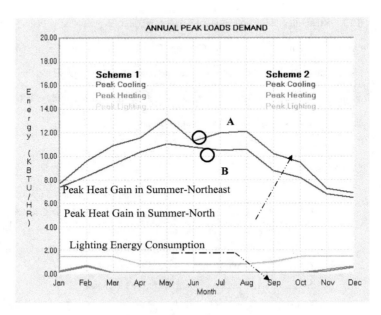

Figure 5- 10 Compare Peak Radiation Heat Gain and Heat Loss-North-Northeast

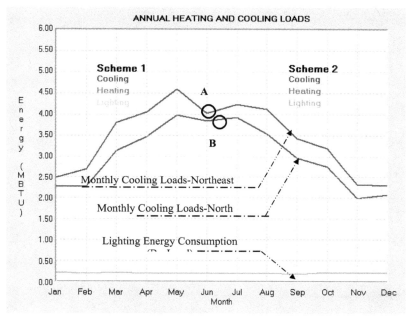

Figure 5- 11.Compare Annual Radiation Heat Gain and Heat Loss-North-Northeast

108

Starting from North, continue from 1 to 6

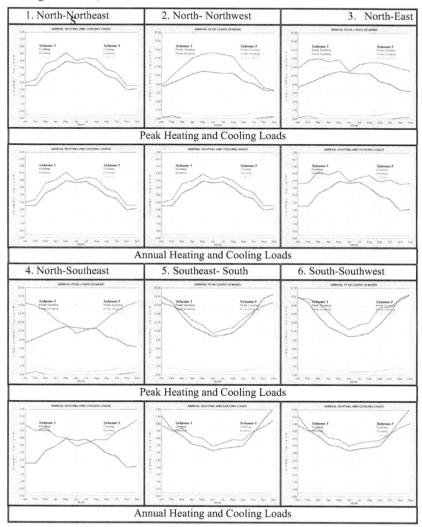

Figure 5- 12.Compare Peak and Annual Heat Gain and Heat Loss-Start from North

110

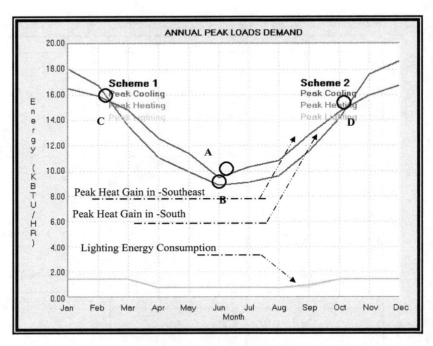

Figure 5- 13 Compare Peak Radiation Heat Gain and Heat Loss-South-Southeast

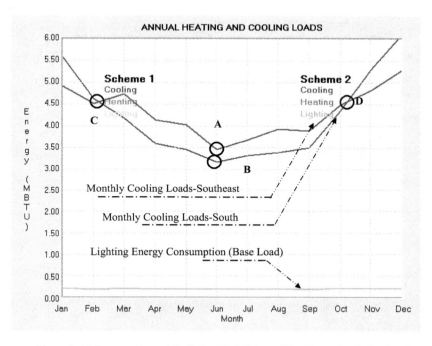

Figure 5- 14 Compare Annual Radiation Heat Gain and Heat Loss--South-Southeast

112

For starting from South, proceed to below

1. South- Southeast	2. South-Southwest	3. South-East

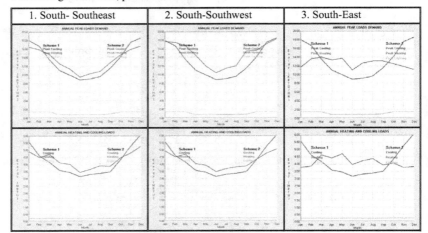

Figure 5- 15.Compare Peak and Annual Heat Gain and Heat Loss-Start from South

5.2.4 Analysis

- North has lower annual heat gain (B) than Northeast (A). The difference between A and B is monthly cooling loads saving.

- South has lower peak summer heat gain (B) than Southeast (A) and better in having free cooling than Southeast (C and D). The difference between A and B is peak cooling loads saving

- South has lower summer heat gain (B) than Southeast (A). The difference between A and B is peak cooling loads saving. South also has better balance in having free cooling (C and D)..

114

116

5.2.5 Solutions. The most desirable site is the site that minimize energy consumption and maximize energy savings By comparing between 2 schemes of several schemes, it can be concluded that South is the most desirable site orientation that has the minimum summer heat gain (cooling loads) and winter heat loss (heating loads), and has the best climate balance between summer heat gain (cooling loads) and winter heat loss (heating loads). Figure 5-16 shows the summary of the results of optimize diagram.

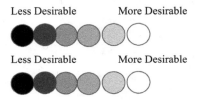

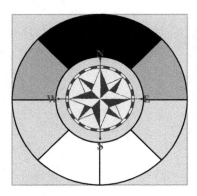

Figure 5- 16 Ranking the Most Desirale Orientation for Site Selection in Bangkok

5.3 Case Scenario 3: Optimize the Use of Percent of Glass

The ideal solutions are to maximize heat gain or minimize heat loss during under-heated periods in winter, and minimize heat gain or maximize heat loss during over-heated period in summer. This case scenario is to analyze maximum or minimum monthly solar insolation characteristics.

5.3.1 Problem definition. Architects have form or shape in mind to design the project in Chicago. They want to know the thermal characteristics of each building façades when facing different orientation so they can decide the percent of glass for window wall used in the façade exposure that allow maximum sun during the under-heated periods and protect to receive minimum sun during the over-heated periods.

5.3.2 Strategies. From each orientation, month, and season the under-heated periods and over-heated period are being analyzed to find the potential optimum percent of glass used in each façade oriented. The followings are the methods to analyze.

- Identify the best and the worst under-heated and over-heated time periods. Time period for over-heated and under-heated area, which show better summer-winter relationship—are more livable. A balance can be found between the over-heated periods when we avoid solar radiation and under-heated periods when we seek radiation. Figure 5-17 to 5-20 shows the under-heated periods and over-heated periods for different orientation. Figure 5-21 to 5-31 shows the optimum percent of glass for different orientation.

- Shading analysis which are based on maximum throughout the year in over-heated times the building need shades (either use appropriate glass types in scenario 4 or shading devices in scenario 5), and in under-heated times the building window wall should receive the sun.

5.3.3 Steps to do

- In the software for both schemes, under climate location pull down box choose Chicago, under orientation choose North under glass type choose monolithic clear glass, choose none for both overhang and fin.
- In the scheme 1 under fenestration choose 80% glass
- In the scheme 2 under fenestration choose 10% glass
- Click the Annual Heat Cool command button to obtain the monthly heating, cooling, and lighting in the graphic screen.
- From the graph, identify the area the under-heated periods and over-heated period occurred.
- Click the Percent Glass command button to obtain the monthly heating, cooling, and lighting through different percent glass.
- Determine the optimum percent of glass for window wall.
- Continue to change to East, South, and West one at a time, and continue to analyze.
- Repeat step 5-7.

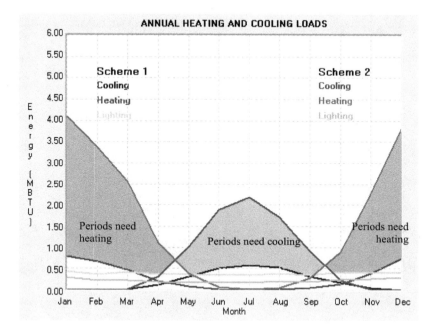

Figure 5- 17. Under-heated periods and over-heated periods—North orientation

120

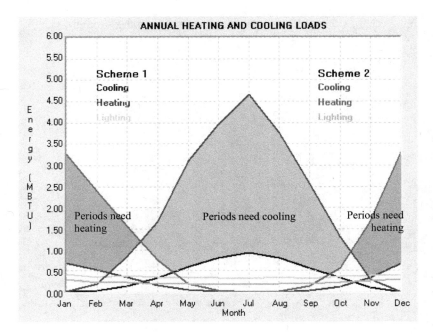

Figure 5- 18. Under-heated periods and over-heated periods—East orientation

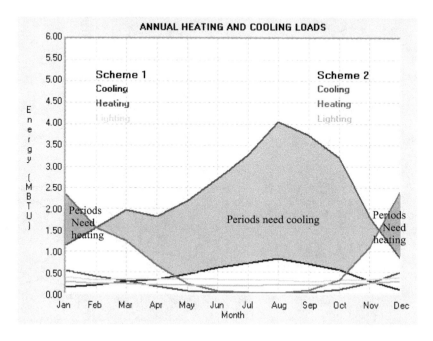

Figure 5- 19. Under-heated periods and over-heated periods—South orientation

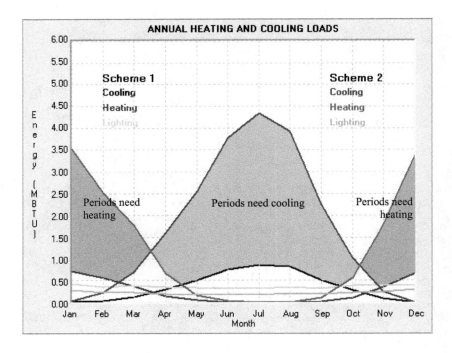

Figure 5- 20. Under-heated periods and over-heated periods—West orientation

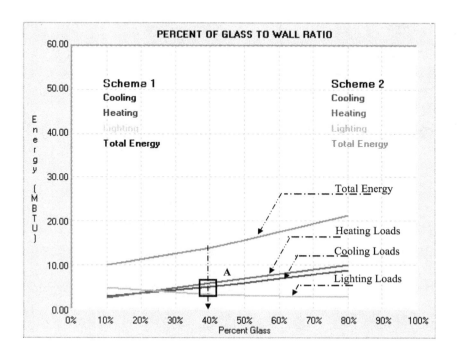

Figure 5- 21.North Facade-Insulated Clear

124

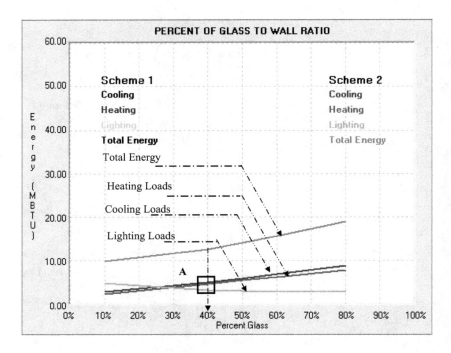

Figure 5- 22. North Facade--LowE Clear

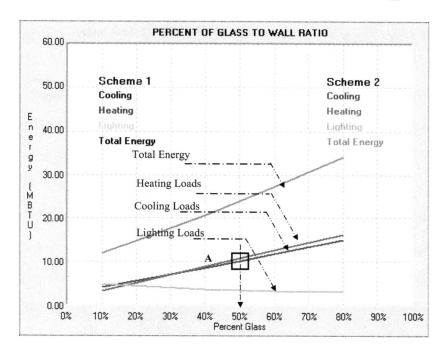

Figure 5- 23.East Façade-Monolithic Tint

126

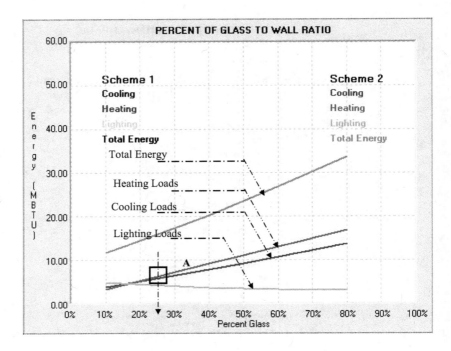

Figure 5- 24. West Façade-Monolithic Tint

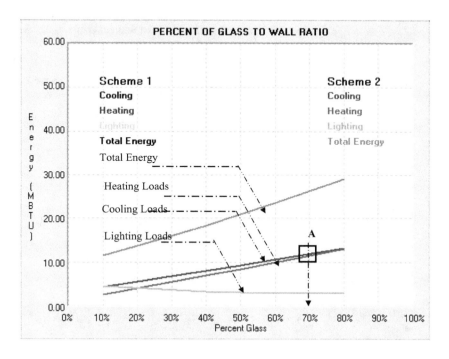

Figure 5- 25. South Façade-Monolithic Tint-0.50 Overhang

128

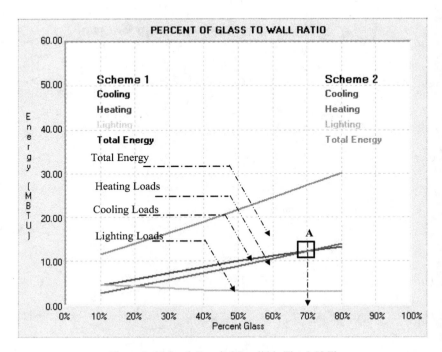

Figure 5- 26.South Façade-Monolithic Tint-0.50 Fin

129

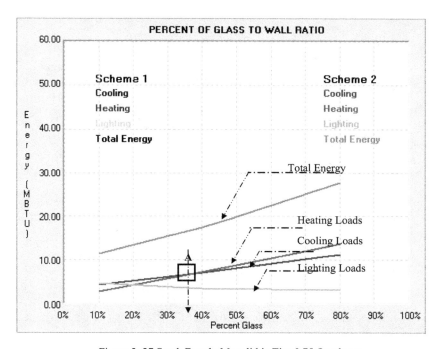

Figure 5- 27.South Façade-Monolithic Tint-0.75 Overhang

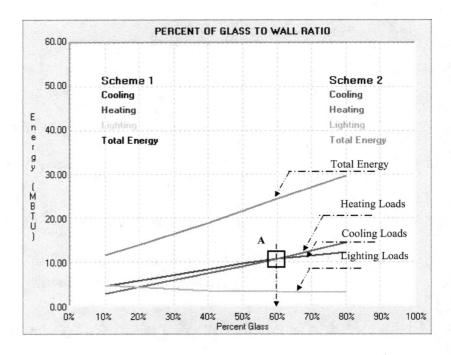

Figure 5- 28.South Façade-Monolithic Tint-0.75 Fin

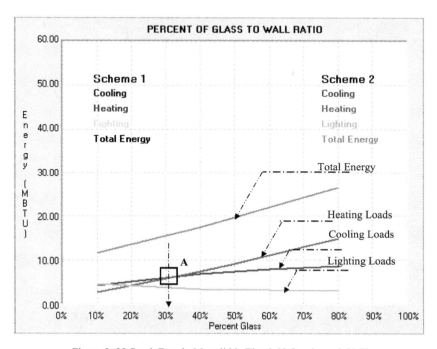

Figure 5- 29.South Façade-Monolithic Tint-0.50 Overhang-0.50 Fin

132

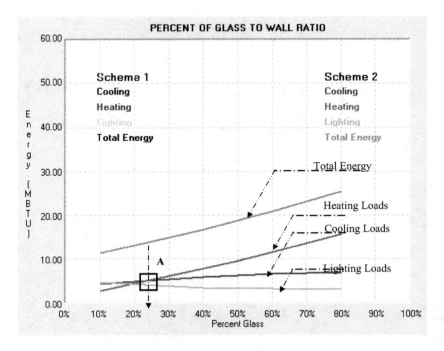

Figure 5- 30.South Façade-Monolithic Tint-0.75 Overhang-0.50 Fin

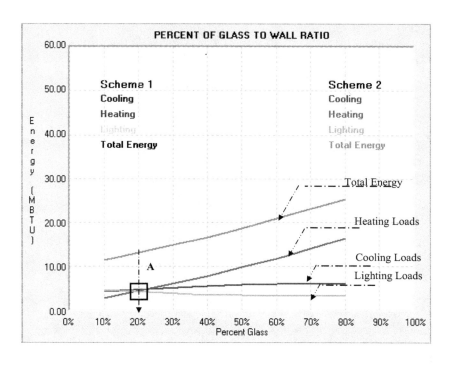

Figure 5- 31.South Façade-Monolithic Tint-0.75 Overhang 0.75 Fin

134

5.3.4 Analysis

- North has worst periods of heat loss in winter and unbalance relationship with heat gain in summer. The façade needs to protect from loosing heat in winter.
- East has worst periods of heat gain in summer and heat loss in winter and unbalance relationship with heat gain and heat loss. In summer the façade need shade or control percent glass while in winter need to control heat loss.
- South has worst periods of heat gain in summer and better balance relationship between summer and winter periods.
- West has worst periods of heat gain in summer and heat loss in winter and unbalance relationship with heat gain and heat loss.
- Figure 5- 21 Optimum for this façade is 40 percent of glass (point A) the point that heating and cooling energy are balance before starting to increase.
- Figure 5- 22 Optimum for this façade is 40 percent of glass (point A) the point that heating and cooling energy are balance before starting to increase or decrease.
- Figure 5- 23 Optimum for this façade is 50 percent of glass (point A) the point that heating and cooling energy are balance before starting to increase.
- Figure 5- 24 Optimum for this façade is 25 percent of glass (point A) the point that heating and cooling energy are balance before starting to increase.
- Figure 5- 25 Optimum for this façade is 70 percent of glass (point A) the point that heating and cooling energy are balance before starting to increase or decrease.
- Figure 5- 26 Optimum for this façade is 70 percent of glass (point A) the point that heating and cooling energy are balance before starting to increase or decrease
- Figure 5- 27 Optimum for this façade is 35 percent of glass (point A) the point that heating and cooling energy are balance before starting to increase or decrease.

- Figure 5- 28 Optimum for this façade is 60 percent of glass (point A) the point that heating and cooling energy are balance before starting to increase or decrease.
- Figure 5- 29 Optimum for this façade is 30 percent of glass (point A) the point that heating and cooling energy are balance before starting to increase or decrease.
- Figure 5- 30 Optimum for this façade is 25 percent of glass (point A) the point that heating and cooling energy are balance before starting to increase or decrease.
- Figure 5- 31 Optimum for this façade is 20 percent of glass (point A) the point that heating and cooling energy are balance before starting to increase or decrease..

5.3.5 Solutions. Percent glass of window wall for the façade should be high enough in order to receiving daylight, view, and thermal comfort. The optimum values are helpful for architects to make the right decisions.

- South orientation can attain the optimum percent of glass of 70% by using the shades like overhang not exceed 0.50 ratio, or fin not exceed o.50 ratio, but not both. This is because South has the worst periods of heat gain in summer but better balance relationship between summer and winter periods.

- North has worst periods of heat loss in winter and unbalance relationship with heat gain in summer. The façade needs to protect from loosing heat in winter. The appropriate percent of glass for this façade is 40 %.

- East has worst periods of heat gain in summer and heat loss in winter, and unbalance relationship with heat gain and heat loss. In summer the façade need shade or control percent glass, while in winter need to control heat loss. The appropriate percent of glass for this façade is 50%.

- West has worst periods of heat gain in summer and heat loss in winter, and unbalance relationship with heat gain and heat loss. In summer the façade need shade or control percent glass, while in winter need to control heat loss. The appropriate percent of glass for this façade is 25 %.

5.4 Case Scenario 4: Select the Glass Types for Building Facades.

The calculation of heat transmission through the use of tabulate data is laborious and time consuming. The graphic system has an advantage over statistical approach. The method shown here can work for any climate location, latitude or orientation that would like to investigate. The heat effects through glass or opaque surfaces can be determined with one reading. Since in architectural problems the irradiation of horizontal and vertical surfaces is of primary interest, the graphs and charts were developed for this exposure. The method can be adapted to any surface plane, and to other condition such as semi-cloudy or cloudy atmosphere.

The façade of the module performs the role as the relative performance of a radiation filter between outdoor and indoor conditions. To contrast a solid wall with a glass window, it is more realistic to compare their role as heat barriers with full thermal performance. The importance of solar control can be shown by compare the amount of heat entering a building through its various components. The relative importance of the

138

components with respect to undesirable heat gains is apparent. The windows account for the greatest amounts of heat entering the building therefore shading them offers the greatest protection. Only a synthesized integration of all components will result in a climate balance shelter.

5.4.1 Problem Definition. The building located in Chicago has the building façades and architects have to select the appropriate glass types used in the façade exposure that allow maximum sun during the under-heated periods and protect to receive minimum sun during the over-heated periods. Find the best glass types to serve the following design criteria:

- Allow maximum sun during the under-heated periods and protect to receive minimum sun during the over-heated periods.
- Maximize daylight penetration through the exposure.
- Maximize electrical light energy savings

5.4.2 Strategies. Since west-facing façade has problems from both over-heated and under-heated periods, the scenario would focus on the study of west facade. First is to use full glass window wall or 80% glass, then start to compare the relative change of each glass types. To compare of the effective glass surface method, the shading coefficient was used as a measure. The shading coefficient is the ratio of the total solar heat gain from the transmitted, absorbed, reradiated energy by the shade and glass combination compared to the total solar heat gain from the transmitted, absorbed, reradiated energy by the single unshaded clear glass.

To evaluate the shading coefficient, the transmitted, absorbed, reradiated energy by the shade and glass combination compared was related to the value of 1.00 as a basic index of unshaded regular double strength window glass. As the flat surface shading glass surfaces (ie.tinted or low-e coated), the incident angle of light transmission value was also evaluated. Study the relative effects on the followings' building elements

- Heat transmission
- Daylight availability
- Electrical lighting energy consumption

Figure 5-32 to 5-37 show heat transmission through glass. Figure 5-38 to 5-39 show daylight availability. Figure 5-40 to 5-41 show electrical lighting energy consumption was reduced from daylight.

5.4.3 Steps to do

- In the software program for both schemes, under climate location pull down box choose Chicago, under orientation choose west, under fenestration choose 80 % glass, choose none for both overhang and fin.
- In the scheme 1: under glass type choose monolithic or single clear glass as a base case.
- In the scheme 2: under glass type choose monolithic tinted glass to compare with the base case.
- Click the Percent Glass command button to view the behavior of heat transmission through different glass types.
- Click the Annual HeatCool command button to view the changing in over-heated and under-heated periods and intensity.
- 6. Click the Annual HeatGain command button to view the monthly heat transmission components during over-heated period
- 7. Click the Annual HeatLoss command button to view the monthly heat transmission components during under-heated period

140

- 8. Click the Heat Gain Components command button to view the categories of heat gain such as glass, wall, lights through different building components
- 9. Click the Heat Loss Components command button to view the categories of heat loss, such as glass, wall, lights, through different building components
- 10. Click the Daylight I luminance command button to view the amount of daylight transmission, is it adequate or inadequate.
- 11. In the scheme 2: under glass type continue to choose insulated clear, insulated tinted, low-e clear, low-e gray, low-e green, low-e reflective.
- 12. Repeat the step 6-10

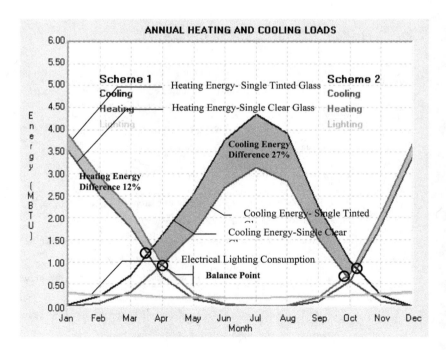

Figure 5- 32. Heat Transmission through Glass

142

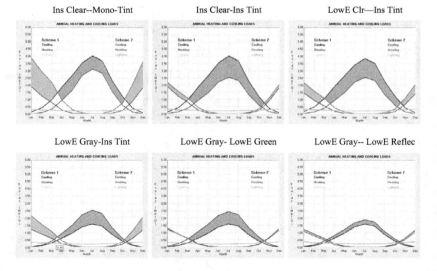

Figure 5- 33.Heat Transmission through Different Glass Types

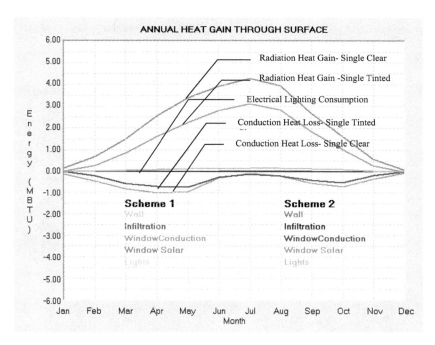

Figure 5- 34.Summer Monthly Heating and Cooling Loads

C

144

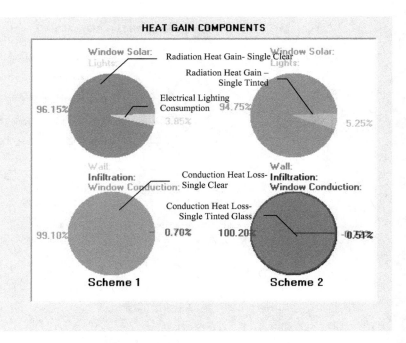

Figure 5- 35.Summer Heating and Cooling Load Components

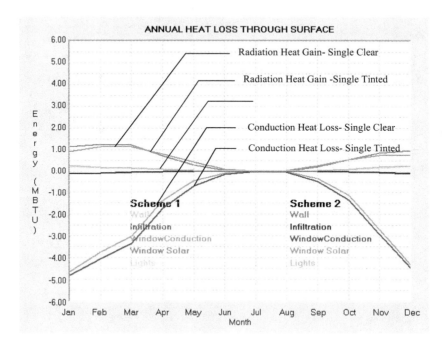

Figure 5- 36.Winter Monthly Heating and Cooling Loads

146

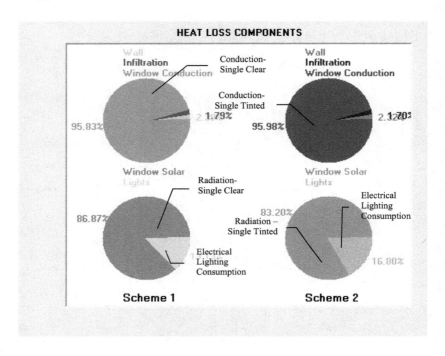

Figure 5- 37.Winter Heating and Cooling Load Components

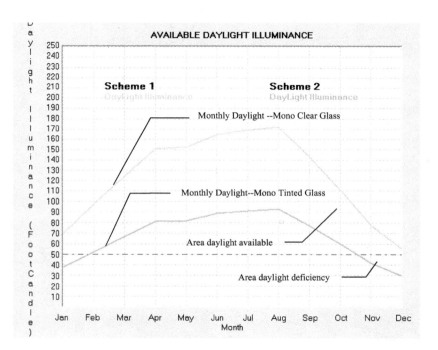

Figure 5- 38..Daylight Availability

148

Mono Clear- Ins Clear

Mono Clear- Ins Tint

Mono Clear- LowE Clear

Mono Clear- LowE Gray

Mono Clear- LowE Green

Mono Clear- LowE Reflec

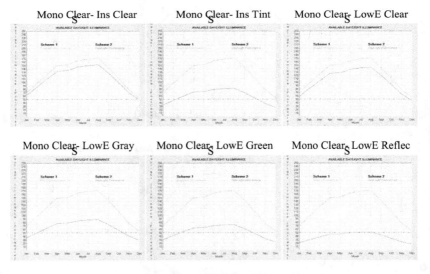

Figure 5- 39. Daylight availability Different Glass Types

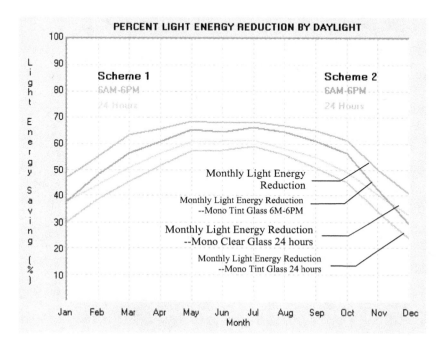

Figure 5- 40.Electrical Lighting Energy Savings

150

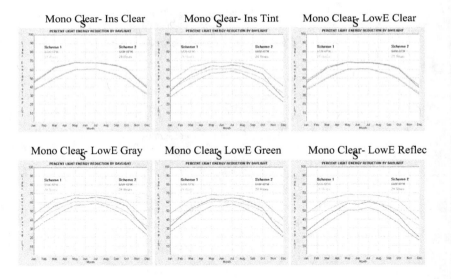

Figure 5- 41.Electrical Lighting Energy Savings of Different Glass Types

5.4.4 Solutions. For heat transmission through the glass surface, proper glass types should support the daylight and heat control. Usually glass deals with over-heated problems by shade protect solar radiation, which has effects on daylight and electrical light consumption, and deals with under-heated problems by glass conductivity, which also depend on the percent or ratio of the glass and wall surface. There is no one solution to this scenario; instead there is more than one solution. One glass type is good at heat control; the other is good at day lighting. Figure 5-42 shows ranking the glass types by shading coefficient and light transmission. Figure 5-43 shows ranking the glass types by cooling and heating energy. Figure 5-44 shows ranking the glass types by daylight performance.

- For balanced heat distribution between high heat gain in summer and low heat loss in winter, the best glass is Insulated Tinted glass.
- For heat transmission through the glass surface and heat control, which deals with over-heated problems by shade protect solar radiation, the best glass type is Low-e Gray glass.
- For heat transmission through the glass surface and heat control, which deals with under-heated problems by glass conductivity, which also depend on the percent or ratio of the glass and wall surface, the best glass is Low-e Reflective glass.
- For maximum exposure to daylight availability and electrical light savings, the best glass is Monolithic Clear glass.

152

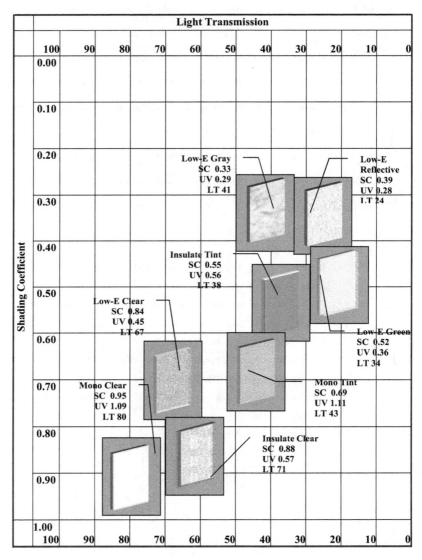

Figure 5- 42. Ranking Glass Types by Shading Coefficient and Light Transmission

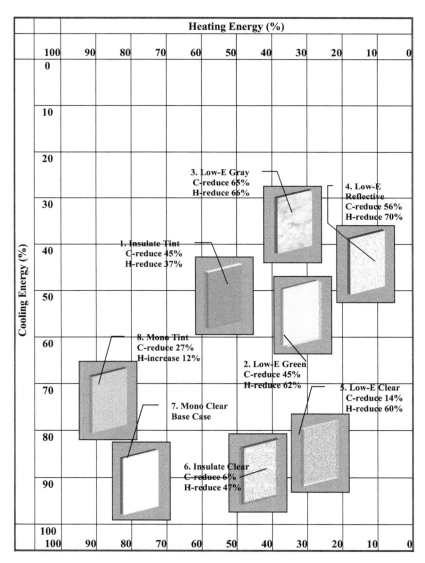

Figure 5- 43.Ranking Glass Types by Cooling Energy and Heating Energy Savings

154

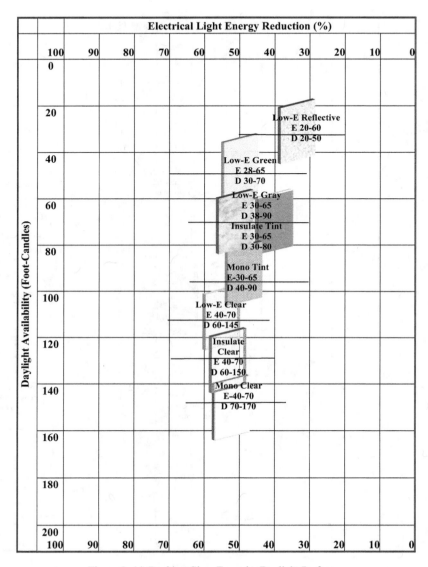

Figure 5- 44. Ranking Glass Types by Daylight Performance

5.5 Case Scenario 5: Shading with Overhang, Fin, and Egg Crate.

To evaluate external shading devices mean to effectively use and place such devices with proper size. This software will help finding proper size and placement to achieve heat balance and energy saving. The size of shade device is related to the value of 1.00. The given size ratio of 0.25, 0.50, 0.75 means the depth of overhang (horizontal) or fin (vertical) is 0.25 x glass height, 0.50 x glass height, and 0.75 x glass height respectively.

5.5.1 Problem definition. The existing building located in Chicago has the building façades that facing West orientation, with 80% window wall. Find the optimum size of shading devices that can control greatest amounts of heat entering the building and have shading capacity for greatest protection. Evaluate the best shading device based on minimize overheated period and maximize balance heat gain and loss. Figure 5-45.shows shading devices problem definition..

5.5.2 Strategies. Method 1: Plot the shade capacity of overhang and fin during over-heated periods for each orientation. Figure 5-46 to 5-61 shows the plot of overhang and fin shade capacity.

- In order to use the shading in the software program, it is necessary first to define times, months and seasons, and then to define the direction-orientation-where shading is needed. Temperature which rise above the **climate balance points** will define as over-heated period, and which rise below the **climate balance points** will define as under-heated period

- Check annual heating and cooling to see the over-heated periods. Check overhang ratio and fin ratio to see the how the overhang and fin can reduce over-heated periods, and how can the shading method yield the best energy saving.

- Make use of the best combination between the shading coefficient from glass and shading devices help solving the heat transmission problems through the facade surface.
- Determine position of the sun when shading is needed
- Determine type and position of the shading device for the overheated periods.
- Evaluate the shading devices.

Method 2: Shading masks: By applied the shading masks to the appropriate shading capabilities, we can find the optimum shading devices for different orientation. Figure 5-62 to 5-69 shows shading mask of the shading devices.

5.5.3 Steps to do

- In the software program for both schemes, under climate location pull down box choose Chicago, under orientation choose north, under glass type choose monolithic clear glass, under overhang and fin choose any ratio except none.
- In the scheme 1, under fenestration choose 80% glass
- In the scheme 2, under fenestration choose 10% glass
- Click the Overhang Ratio command button to obtain the monthly heating, cooling, and lighting that are impacted from different overhang size.
- Click the Fin Ratio command button to obtain the monthly heating, cooling, and lighting that are impacted from different fin size.
- In the scheme 2, under orientation continue to choose northeast, east, southeast, south, southwest, west, northwest.
- Repeat step 4-5

To determine of the **shading masks** from the selected shading devices received from the above steps, do the following;

- In the software program for both schemes, under climate location choose Chicago, under orientation choose north, under fenestration choose 80% glass, under glass type choose monolithic clear glass.
- In the scheme 1under overhang choose none and under fin choose none.
- In the scheme 2 under overhang choose 0.50 and under fin choose none.
- Click the Month HeatCool command button to obtain the shading mask resulted from the selected overhang size.
- In the scheme 1 under overhang choose none and under fin choose none.
- In the scheme 2 under overhang choose 0.75; under fin choose none.
- Click the Month HeatCool command button to obtain the shading mask resulted from the selected overhang size.
- In the scheme 1 under overhang choose none and under fin choose none.
- In the scheme 2 under overhang choose none and under fin choose 0.25.
- Click the Month HeatCool command button to obtain the shading mask resulted from the selected overhang size.
- In the scheme 1, under overhang choose none, under fin choose none
- In the scheme 2, under overhang choose 0.50, under fin choose 0.25 (Egg crate)
- Click the Month HeatCool command button to obtain the shading mask resulted from the selected overhang size.
- In the scheme 1 under overhang choose none and under fin choose none.
- In the scheme 2 under overhang choose 0.75 and under fin choose 0.25 (Egg crate)
- Click the Month HeatCool command button to obtain the shading masks resulted from the selected overhang size.
- Continue on both schemes, under orientation continue to choose East, South, and West

158

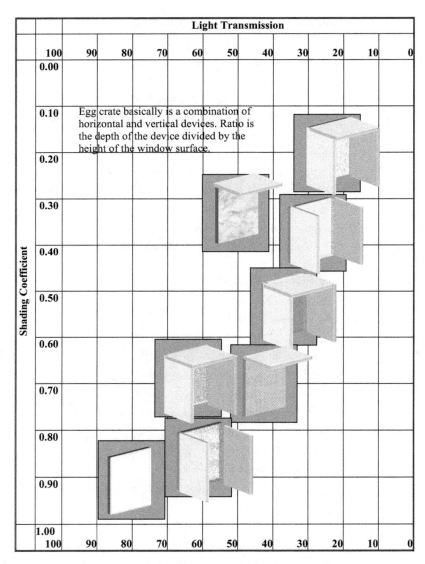

The chart shows Light Transmission (100 to 0) on the horizontal axis and Shading Coefficient (0.00 to 1.00) on the vertical axis.

Egg crate basically is a combination of horizontal and vertical devices. Ratio is the depth of the device divided by the height of the window surface.

Figure 5- 45.Shading Devices and Their Relationship to Glass Types

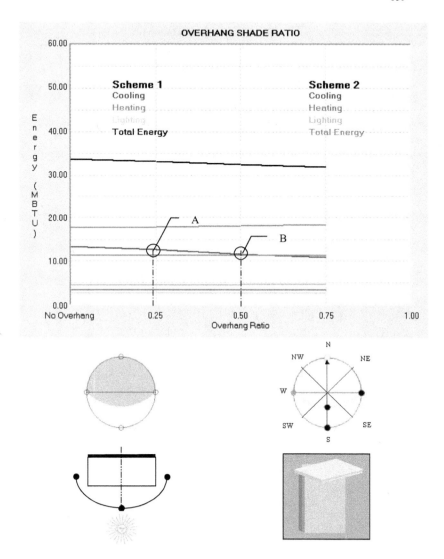

Figure 5- 46.Plot the Overhang Shade Capacity—North Orientation

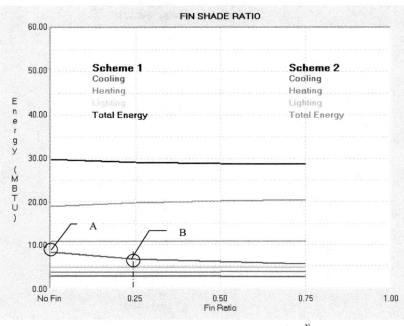

FIN SHADE RATIO

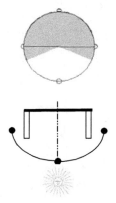

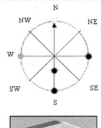

Figure 5- 47.Plot the Fin Shade Capacity—North Orientation

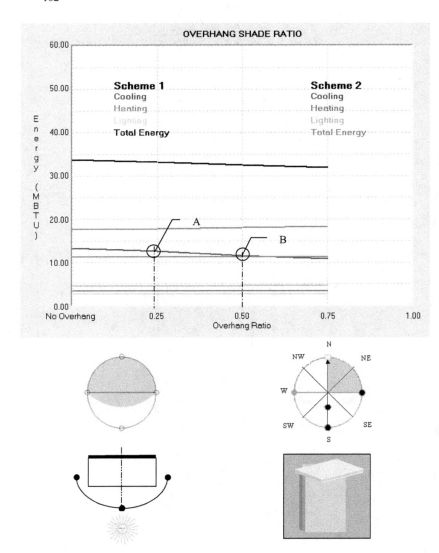

Figure 5- 48.Plot the Overhang Shade Capacity—Northeast Orientation

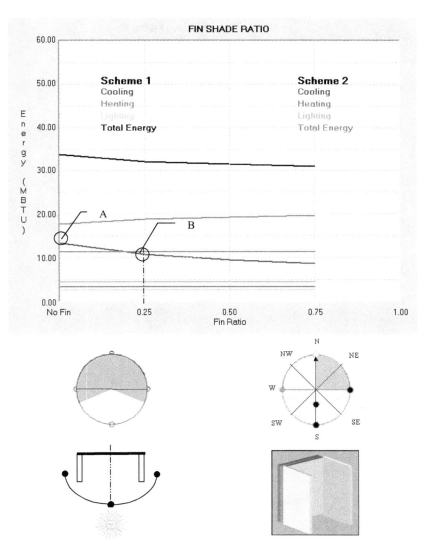

Figure 5- 49.Plot the Fin Shade Capacity—Northeast Orientation

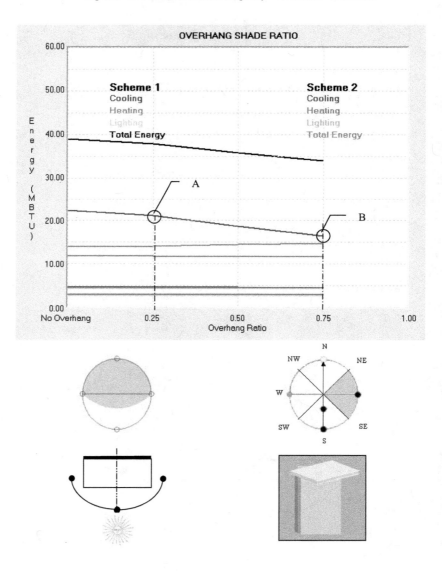

Figure 5- 50.Plot the Overhang Shade Capacity—East Orientation

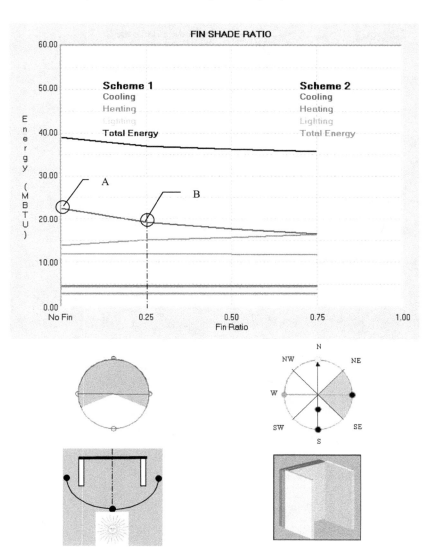

Figure 5- 51.Plot the Fin Shade Capacity—East Orientation

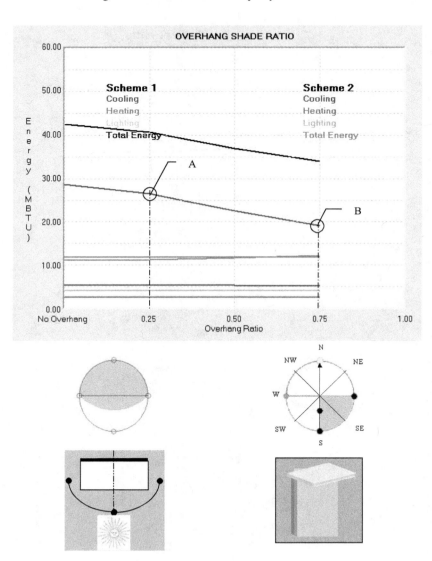

Figure 5- 52.Plot the Overhang Shade Capacity—Southeast Orientation

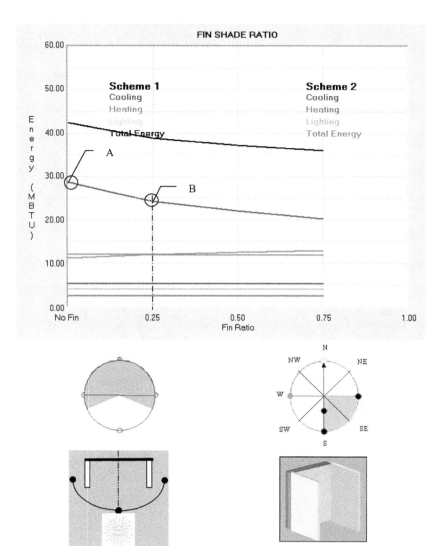

Figure 5- 53.Plot the Fin Shade Capacity—Southeast Orientation

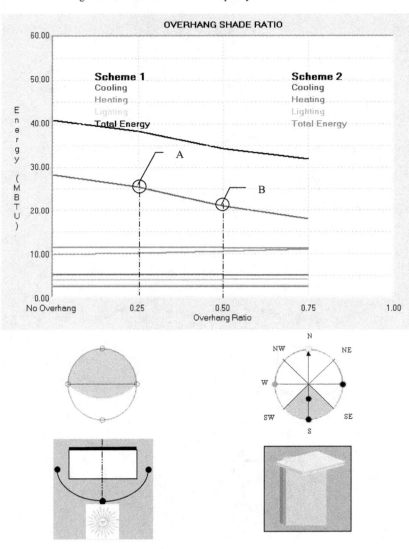

Figure 5- 54.Plot the Overhang Shade Capacity—South Orientation

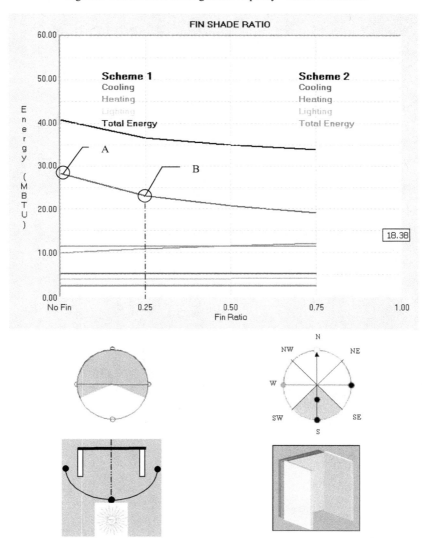

Figure 5- 55.Plot the Fin Shade Capacity—South Orientation

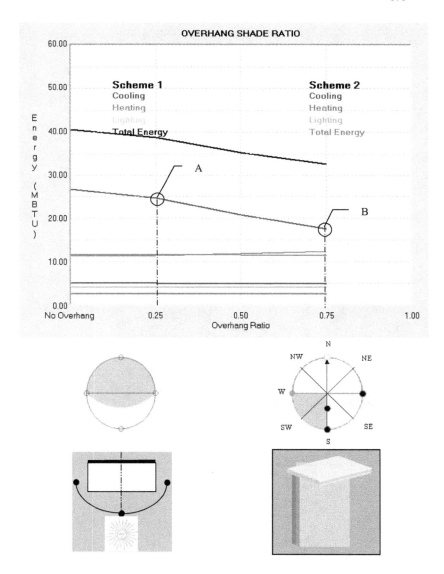

Figure 5- 56.Plot the Overhang Shade Capacity—Southwest Orientation

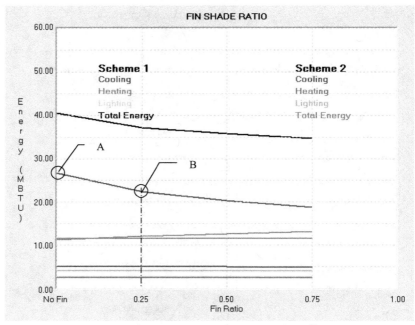

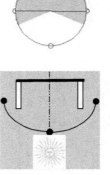

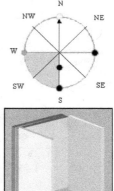

Figure 5- 57.Plot the Fin Shade Capacity—Southwest Orientation

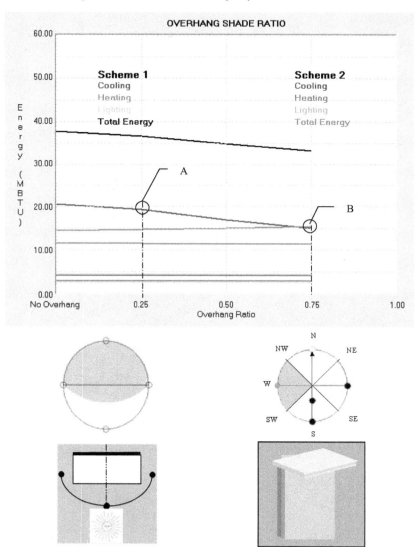

174

Figure 5- 58.Plot the Overhang Shade Capacity—West Orientation

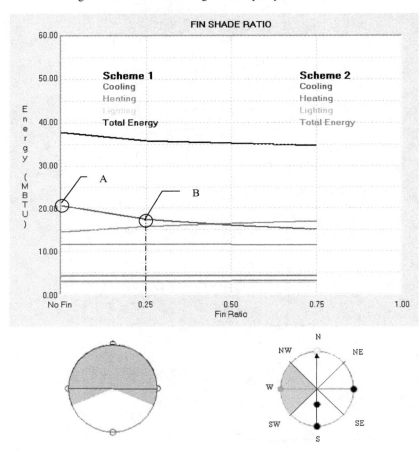

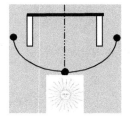

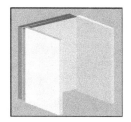

176

Figure 5- 59. Plot the Fin Shade Capacity—West Orientation

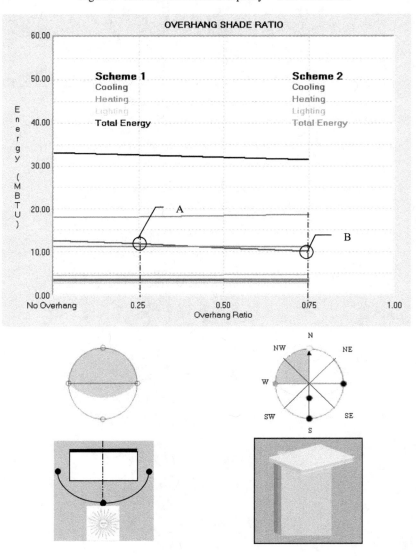

Figure 5- 60.Plot the Overhang Shade Capacity—Northwest Orientation

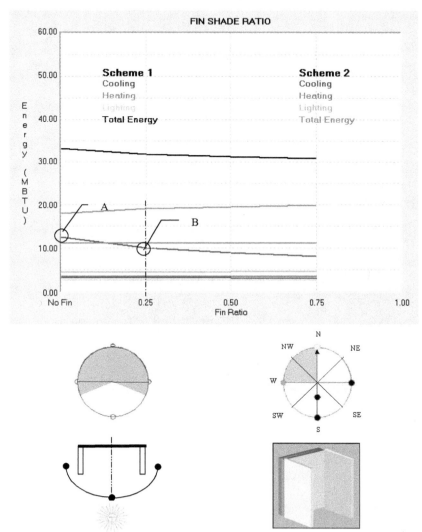

Figure 5- 61.Plot the Fin Shade Capacity—Northwest Orientation

North

–0.5 Ovhgs
--0.5 Ovhg 0.25 Fin
--0.25 Fin
–0.75 Ovhgs
--0.75 Ovhg 0.25 Fin
ANNUAL HEATING AND COOLING LOADS
Scheme 1
Scheme 2

Figure 5- 62.Shading Mask—North Orientation

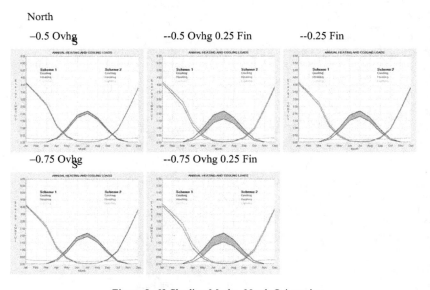

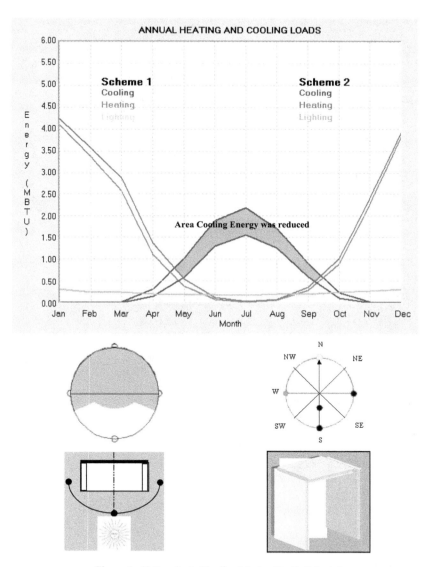

Figure 5- 63.Egg Crate Shading Mask—North Orientation

180

East

–0.5 Ovhg   --0.5 Ovhg 0.25 Fin   --0.25 Fin

–0.75 Ovhg   --0.75 Ovhg 0.25 Fin

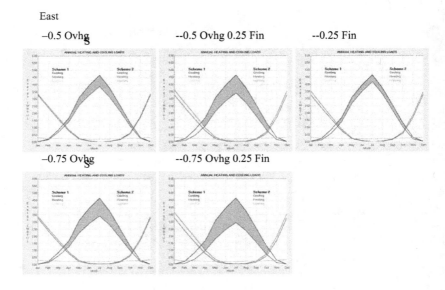

Figure 5- 64.Shading Mask—East Orientation

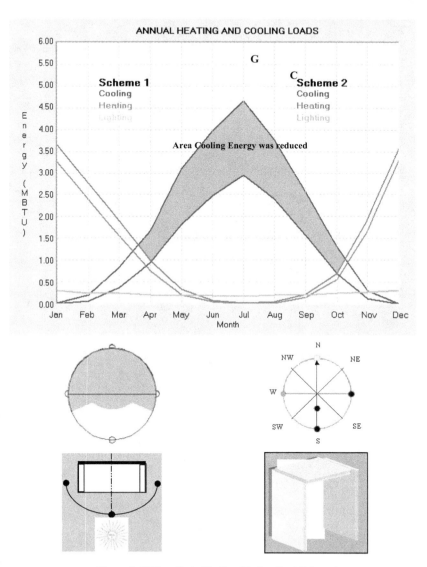

Figure 5- 65.Egg Crate Shading Mask—East Orientation

182

South

–0.5 Ovhg   --0.5 Ovhg 0.25 Fin   --0.25 Fin

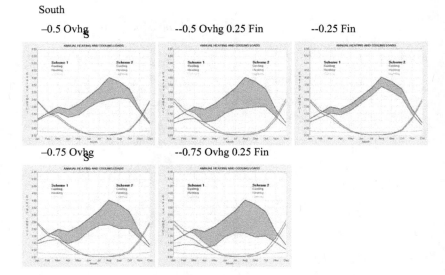

–0.75 Ovhg   --0.75 Ovhg 0.25 Fin

Figure 5- 66.Shading Mask—South Orientation

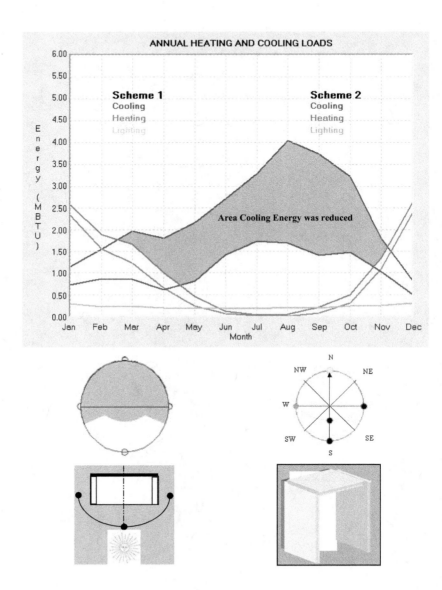

ANNUAL HEATING AND COOLING LOADS

Scheme 1
Cooling
Heating
Lighting

Scheme 2
Cooling
Heating
Lighting

Area Cooling Energy was reduced

Figure 5- 67 Egg Crate Shading Mask—South Orientation

West

–0.5 Ovhg   --0.5 Ovhg 0.25 Fin   --0.25 Fin

–0.75 Ovhg   --0.75 Ovhg 0.25 Fin

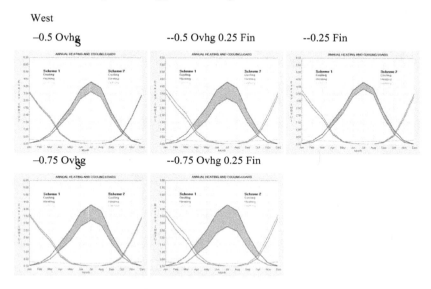

Figure 5- 68.Shading Mask—West Orientation

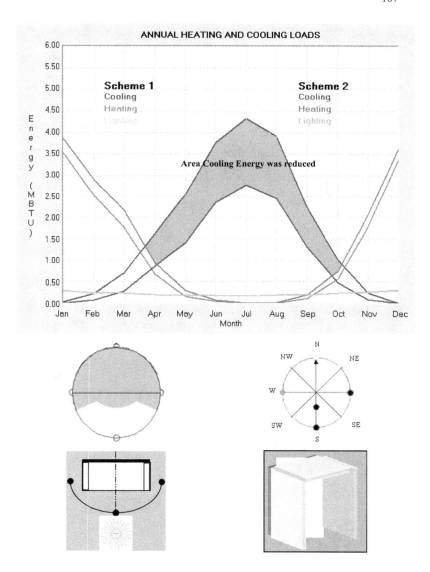

Figure 5- 69.Egg Crate Shading Mask—West Orientation

5.5.6 Solutions. From each different orientation of the façade shading mask during over-heated periods in summer, find the shading devices (overhang, fin, and egg crate) that are able to shade maximum over-heated area according to the orientation, periods, and seasons.

From the above, the optimum sizes of shading devices are 0.5 overhangs, 0.75 overhangs, and 0.25 fins. By continue applied the size to the design selection, it can be determined the appropriate shading masks for all orientation.

- For North orientation the best shading device is 0.75-0.25 Egg Crate

- For East orientation the best shading device is 0.75-0.25 Egg Crate

- For South orientation the best shading device is 0.75-0.25 Egg Crate

- For West orientation the best shading device is 0.75-0.25 Egg Crate

CHAPTER 6

CONCLUSION: RESEARCH FINDINGS

This research provides the finding of the software tool called "Simplified Energy Modeler". This software tool has been developed to give the preliminary prediction for the energy performance of the building façade according to the design parameters: Two different skin facades can be compared according to the climate location of Chicago and Bangkok, orientation, window to wall ratio, glass types, horizontal (overhang) and vertical (fin) shades.

6.1 Contribution

This research makes several contributions to architects and those who use this software tool, will enhance to the practice of architecture and energy efficiency that will lead to sustainability as the following:

1. Enhance the design of environmental responsive projects; ability to adapt design projects to the climate region where the building is situated.

2. Enable to compare energy conservation and architecture parameters information and choose better alternatives among the design schemes.

3. Better understand the relationship between energy design parameters and architectural elements.

4. Enable to choose types of energy efficiency and high performance output results that appropriate.

5. Support architects in the sustainable design and environmental responsive decision making during the design process.

6.2 Implement the Software as an Educational Tool

The real challenge is to make this software tool as an educational tool applicable to the earliest stage of design, where more informed analysis of possible alternatives could yield the most benefit and the greatest cost savings both socio-economic and environmental. "There is nothing more essential in moving toward the long term goal of sustainability than teaching the next generation of architects, scientist and engineers how to incorporate sustainability to their works. (Environmental Protection Agency's Assistant Administrator for Research and Development Dr. Paul Gilman)

This software program tool can be used and integrated in the architectural curriculum core courses, design studio, and as the distance learning educational tool to build an understanding of sustainability principle by experiment with data. By installing this software into the university server and become an Internet environment for public to access to this software or download from the web to personal computers. This program can become available as freeware or shareware in public domain in the university.

Educators and their students will be able to share experience the different selection of energy efficient uses in building responsive guidelines for sustainable solutions based on the criteria, data sets, models, and databases they are aware of. The learning period is approximately half an hour but the learning from experiment is tremendous in term of impacts of their design on sustainability. The documentation required is the built-in help in the program. Figure 6-1 shows the platform and learning curve of the software.

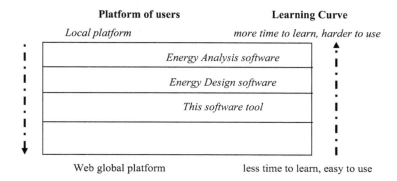

Figure 6- 1 Platform and Learning Curve

6.3 Future Research:

This research would be even more useful if future research will be explore to further ideas and concept of the following studies:

6.3.1 Interactive Data Edit Ability. Once architects create the geometric model, the internal space and zones will be modified from the recorded location of actual weather data. Building schedule and occupancy patterns will be changed in the process to insure proper functional requirements. By using this future program, the models can be evaluated for energy performance, and can be compared each time the various elements of the schematic design are changed.

6.3.2 Visualization Interface of the design. The creation of the simulation model should assist architects in formalize ideas and test them against a range of design constraints. (Akin, O., 1978.) Therefore the simulation model must represent building form and assist in the visualization of the design. This calls for a very geometric approach to representation. The challenge in this approach is to produce an interface within which geometric modeling is as simple and flexible as a sketch, yet can be used for detailed analysis at later stages in the design development. Just as a sketch develops and becomes more sophisticated, so must the geometric model and its analysis. A sketch can focus on a very small area of a model or only one aspect; this can be true for simulation tool as well.

6.3.3 Input Data Utilization. In order to make, a future design tool to be used at the earliest stage in the building design process, the separation between design and analysis tools must be overcome. The primary cause of this separation is the detailed nature and amount of input required to describe the building model. Accommodating these requirements requires a very flexible interface that satisfies the followings:

- Reduces the perceived input requirements to define a model.
- Maximizes utilization of whatever information is input.
- Allows constant development and refinement of the model.

Reducing input requirements and making maximum use of input data require more than traditional geometric interfaces that focus on geometric entities. Basic relationships between architectural elements can be used to create geometric relationships between objects that can significantly reduce the time required when inputting and editing the model.

6.3.4 Model Geometry Automatically Updated. When working with objects that are geometrically related to others, the plane equation is already defined, respectively to the 3D cursor plane. Such nodes will only move within the plane of the parent object. [Wall, Window, Door] Interior Wall, Roof; that is entities is locked into movement within plane of their parent object. No nodes are allowed to move outside the defining 3D polygon of the parent.

Defining automatic geometric relationship between building elements can reduce data entry time and substantially increase model edit ability. Basically it means deriving the geometry of one element from the geometry of another by storing the rules used. If the parent element moves, the geometry of the child objects is automatically updated. If implemented correctly, moving part of the floor plan or adjusting its height should automatically update the walls and ceiling of a space. These relationships also extend to child objects such as windows and doors. Any movement of a parent object must automatically update the position of the child objects.

The manipulation of the nodes (spaces and zones) is important part of the interface as it allows for increased flexibility in what geometry can be modeled. The architects or designers can also take advantage of the efficiency offered by a parametric command.

194

6.3.5 Progressive Data Input. Most existing energy analysis tools provide very visual feedback during calculations. This means that the process being undertaken is hidden from the user. Mistakes in modeling that are not immediately visually apparent must be determined from a close examination of any input or output. A more appropriate implementation is to structure calculations around a full set of basic assumptions and default values, any of which the user can change at any time. Progressive data input refers to the ability to enter the subset of data required to model a particular process and generate quick results. As the design gradually resolves, more detailed information is added to the model, making the results progressively more accurate. This makes the modeling process more responsive.

6.3.6 Intelligent User Interface A great deal of information can be inferred from the context in which an object is created, or from the group of actions that may have preceded it. It is possible to provide a number of alternate means of invoking the same action, each with different consequences for associated data. In this way, the user need only select a different icon or menu item in order to enter the same initial values, but with a different set of defaults. For advanced users, the ability to change the default inferences may be appropriate.

There are some fundamental elements common to most architectural design, such as floors, walls, ceilings, roofs, windows, doors or shading elements. The correct assignment of elements must be left to the architects' or designers' discretion. Traditional CAD applications concentrate on the drawing process rather than modeling. In a modeling tool, knowledge as to the function of an element is essential. Such information facilitates the scheduling, consumption and cost tracking components, automatic generation of the building characteristics (percentage glazing to surface area, space and zone interrelationship, light or equipment energy per square feet). A modeling tool allows analysis and simulation engines to react differently to different elements.

Given a defined geometry, information such as the surface area of walls, the volume of spaces and areas of intersection between the zones can to be extracted. It is also very crucial to model the day lighting. If the 3D modeling interface can be made sufficiently intuitive, inputting the geometry of the building can greatly reduce the time to gather a significant amount of data input. *David Bartholemew in1997 described this as "the validity defined as usefulness"* and suggests that it is not necessary for a model to capture natural processes particularly well in order to be useful. What is important is the practical application and relevance of its output.

It is argued here that there must to be at least two levels of modeling and analysis: Early in the process, it is necessary to provide interactive design feedback, for testing the viability of an idea, and comparing multiple options and making the preliminary estimation of element sizes. The absolute accuracy requirement at this level is quite low. What is important is its relative accuracy, being able to immediately assess changes resulting from a particular set of design decisions compared to the original conditions. This acts to guide the decision making process in the right direction.

196

The second level of analysis requires a more detailed and comprehensive model and is more likely to be based on fundamental physical processed. In this case absolute accuracy is important. This is the level where the conceptual design tool is essential.

6.3.7 Components and Library Database. The exchange of electronic data is becoming more important. The ability to exchange ideas and design information with other architects and consultants is an essential part of any practice. To facilitate this, there must be a way to encapsulate all the data required to describe a building model so that the other architects and consultants are not disadvantaged by not having the same weather files or material libraries.

An attempt has been made to include all of the information that describes the model and its analysis in the one file. The file is structured in a series that has a header and contains information relating to building geometry, space, massing and form, material specification, climatic data and calculation results.

Process of Electronic Data include

- 1. Get data
- 2. Lookup and sort data
- 3. Display data in the form
- 4. Process data
- 5. Write data
- 6. Distribute data to architects

6.3.8 Interoperability. Operation ability with other tools means providing some basic interface with other applications. The most usual form of this is the output data stored in a file format that can be read by other applications, such as Visual BASIC to transfer the data between DOE2 and AutoCAD or other tools.

6.3.9 Integration of the Tools. There have been a number of attempts at some level of integration within these diverse areas of building performance. The computer software suite for the thermal design of building is a good example of integrated climatic analysis with a solar geometry and thermal analysis engine. The followings are the characteristics of the existing tools;

- Tool explains the impact on building energy use of changes in each set of the parameters
- Tool explains the impact on building energy use from the interrelationship between all of the parameters, when they are changing together.

6.3.10 Material Library Operability The inclusion of a material library within application makes provision for specific manufacturer data to be included. At the more complex level, it allows manufacturers to specify the dimensions of their own library objects. This means that a library containing a range of aluminum curtain walls could be provided that has accurate performance specifications for each element based on actual glass frame ratio and production costs.

APPENDIX A

DEFINITIONS OF SUSTAINABILITY

Definition of Sustainability

1.   "...A method of harvesting or using a resource so that the resource is not depleted or permanently damaged."

2.   "...A lifestyle involving the use of sustainable methods."[3]

3.   "...– "Development that meets the needs of the present generation without compromising the ability of future generations to meet their own needs."[4]

Aesthetic in Sustainability

1.   "...A branch of philosophy dealing with the nature of beauty, art, and taste and with the creation and appreciation of beauty."

2.   "...A particular theory or conception of beauty or art: a particular taste for or approach to what is pleasing to the senses and especially sight."

3. "...A pleasing appearance or effect."[5]

Ethic in Sustainability

1. "...The discipline dealing with what is good and bad and with moral duty and obligation."

2. "...A set of moral principles or values; a theory or system of moral values; the principles of conduct governing an individual or a group; a guiding philosophy."

---

[3] Source: www.m-w.com/cgi-bin/ dictionary (Merriam-Webster online)

[4] United Nations' World Commission on Environment and Development – The Brundtland Commission
[5] Source: www.m-w.com/cgi-bin/ dictionary (Merriam-Webster online)

APPENDIX B

EXAMPLE OF DOE 2.1E INPUT BDL FILE

202

```
*    1 * INPUT LOADS ..
```

L D L   P R O C E S S O R   I N P U T   D A T A

10/05/2004   14:06:09 LDL RUN   1

```
*    2 * RUN-PERIOD    JAN  1, 2001 THRU DEC 31, 2001 ..

*    3 * PARAMETER

*    4 * LAT = 13.92

*    5 * LONG = -100.60

*    6 * ALT = 39

*    7 * T-Z = -7

*    8 * AZI = 0

*    9 * WINHGT = 1.40

*   10 * INF-ACH = 0.052

*   11 * GLASS-NP = 1

*   12 * GLASS-SC = 0.95

*   13 * GLASS-UV = 1.09

*   14 * GLASS-LT = 0.80

*   15 * OVHG-A = 0

*   16 * OVHG-B = 0

*   17 * OVHG-W = 0

*   18 * OVHG-D = 0

*   19 * OVHG-ANGLE = 90

*   20 * LFIN-A = 0

*   21 * LFIN-B = 0

*   22 * LFIN-H = 0
```

```
*   23 *  LFIN-D = 0

*   24 *  RFIN-A = 0

*   25 *  RFIN-B = 0

*   26 *  RFIN-H = 0

*   27 *  RFIN-D = 0

*   28 *  SENSOR = 5

*   29 *  ..

*   30 *  BUILDING-LOCATION

*   31 *  LATITUDE = LAT

*   32 *     LONGITUDE = LONG

*   33 *       ALTITUDE = ALT

*   34 *       AZIMUTH = AZI

*   35 *       TIME-ZONE = T-Z

*   36 *       DAYLIGHT-SAVINGS = YES

*   37 *       HOLIDAY = YES ..

*   38 *  SCH-OFC-LTG = SCHEDULE    THRU DEC 31

*   39 *    (WD)  (1,24) = (0.05 0.05 0.05 0.05 0.05 0.10 0.35 0.50
0.90 0.90 0.90 0.90

*   40 *                 0.90 0.90 0.90 0.90 0.90 0.90 0.90 0.50
0.35 0.35 0.10 0.05)

*   41 *    (SAT) (1,24) = (0.05 0.05 0.05 0.05 0.05 0.05 0.10 0.10
0.50 0.50 0.50 0.50

*   42 *                 0.50 0.50 0.50 0.50 0.10 0.10 0.05 0.05
0.05 0.05 0.05 0.05)

*   43 *    (SUN) (1,24) = (0.05)

*   44 *    (HOL) (1,24) = (0.05)

*   45 *  ..

*   46 *  SCH-INFLTR-WNDW = SCHEDULE
```

```
*   47  *         THRU FEB 28   (ALL)  (1,24)=(1.0)

*   48  *         THRU APR 30   (ALL)  (1,24)=(0.5)

*   49  *         THRU OCT 31   (ALL)  (1,24)=(0.0)

*   50  *         THRU NOV 30   (ALL)  (1,24)=(0.5)

*   51  *         THRU DEC 31   (ALL)  (1,24)=(1.0)   ..

*   52  *  SCH-DAYREP-ON = SCHEDULE   THRU DEC 31
(ALL)(1,5)(0)(6,18)(1)(19,24)(0)  ..

*   53  *  MODULE-WALL = CONSTRUCTION    U-VALUE=0.08 ..

*   54  *  MODULE-PARTITION = CONSTRUCTION    U-VALUE = 0.5 ..

*   55  *  MODULE-GLASS = GLASS-TYPE

*   56  *       PANES = GLASS-NP

*   57  *       SHADING-COEF = GLASS-SC

*   58  *       GLASS-CONDUCTANCE = GLASS-UV

*   59  *       VIS-TRANS = GLASS-LT   ..

*   60  *  MODULE-SP  = SPACE-CONDITIONS

*   61  *        TEMPERATURE = (75)

*   62  *        LIGHTING-W/SQFT = 1.2

*   63  *        LIGHT-TO-SPACE = 0.8

*   64  *        LIGHTING-SCHEDULE = SCH-OFC-LTG

*   65  *        INF-SCHEDULE = SCH-INFLTR-WNDW

*   66  *        INF-METHOD = AIR-CHANGE

*   67  *        AIR-CHANGES/HR = INF-ACH

*   68  *        ZONE-TYPE = CONDITIONED

*   69  *        LIGHTING-TYPE=REC-FLUOR-NV   ..

*   70  *  SET-DEFAULT FOR SPACE

*   71  *        DAYLIGHTING = YES

*   72  *        ZONE-FRACTION1 = 1

*   73  *        LIGHT-SET-POINT1 = 50
```

```
*  74 *        LIGHT-CTRL-TYPE1 = CONTINUOUS

*  75 *        LIGHT-REF-POINT1 = (7.5,SENSOR,3)

*  76 *        DAYLIGHT-REP-SCH = SCH-DAYREP-ON  ..

*  77 * SET-DEFAULT FOR EXTERIOR-WALL

*  78 *        CONSTRUCTION = MODULE-WALL

*  79 *        HEIGHT = 14  ..

*  80 * SET-DEFAULT FOR WINDOW

*  81 *        HEIGHT = WINHGT

*  82 *        GLASS-TYPE = MODULE-GLASS

*  83 *        OVERHANG-A= OVHG-A

*  84 *        OVERHANG-B = OVHG-B

*  85 *        OVERHANG-W = OVHG-W

*  86 *        OVERHANG-D = OVHG-D

*  87 *        OVERHANG-ANGLE= OVHG-ANGLE

*  88 *        LEFT-FIN-A = LFIN-A

*  89 *        LEFT-FIN-B = LFIN-B

*  90 *        LEFT-FIN-H = LFIN-H

*  91 *        LEFT-FIN-D = LFIN-D

*  92 *        RIGHT-FIN-A = RFIN-A

*  93 *        RIGHT-FIN-B = RFIN-B

*  94 *        RIGHT-FIN-H = RFIN-H

*  95 *        RIGHT-FIN-D = RFIN-D  ..

*  96 * SET-DEFAULT FOR  INTERIOR-WALL

*  97 *        CONSTRUCTION = MODULE-PARTITION   ..

*  98 * MODULE   = SPACE

*  99 *        AREA = 225   VOLUME = 2250

* 100 *        SPACE-CONDITIONS = MODULE-SP  ..

* 101 * MODULE-WL = EXTERIOR-WALL
```

```
*  102  *          WIDTH = 15
*  103  *          TILT=90
*  104  *           AZIMUTH=0   ..
*  105  * MODULE-WIN = WINDOW
*  106  *           WIDTH = 15 ..
*  107  * LEFTPART =I-W
*  108  *        AREA=150
*  109  *        INT-WALL-TYPE = ADIABATIC ..
*  110  * RIGHTPART=I-W       LIKE LEFTPART   ..
*  111  * BACKPART =I-W       LIKE LEFTPART   ..
*  112  * INTFLR  = I-W
*  113  * AREA=225
*  114  * INT-WALL-TYPE = ADIABATIC
*  115  * TILT=180  ..
*  116  * INTCEIL = I-W
*  117  * AREA=225
*  118  * INT-WALL-TYPE = ADIABATIC
*  119  * TILT=0     ..
*  120  * LOADS-REPORT
*  121  * SUMMARY = (LS-D,LS-F,LS-G) ..
*  122  * END ..
*  123  * COMPUTE LOADS  ..
*  124  * SAVE-FILES  ..
*  125  * STOP ..
```

APPENDIX C

EXAMPLE OF DOE2.1E VERIFICATION FILE

Source : DOE-2 Program

REPORT- **LV-A** GENERAL PROJECT AND BUILDING INPUT

PERIOD OF STUDY

STARTING DATE	ENDING DATE	NUMBER OF DAYS
1 JAN 2003	31 DEC 2003	365

SITE CHARACTERISTIC DATA

BUILDING

STATION	LATITUDE (DEG)	LONGITUDE (DEG)	ALTITUDE (FT)	TIME ZONE	AZIMUTH (DEG)
NAME					
CHICAGO, IL	42.0	88.0	610.	6 CST	135.0

WEATHER FILE- CHICAGO, IL

REPORT- **LV-B** SUMMARY OF SPACES OCCURRING IN THE PROJECT

NUMBER OF SPACES	1	EQUIP	EXTERIOR 1	INTERIOR 0	

SPACE	LIGHTING SPACE*FLOOR (WATT / SQFT)	EQUIP SPACE*FLOOR (WATT / SQFT)	PEOPLE	INFILTRATION (WATT / SQFT)	AIR CHANGES METHOD	AIR-CHANGE
MODULE	TYPE EXT	AZIMUTH				
MODULE	1.0	1.20	0.0	0.00	PER HOUR	0.08

BUILDING TOTALS 0.0 225.00 2025.00

	AREA (SQFT)	VOLUME (CUFT)
	225.00	2025.00

WEATHER FILE- CHICAGO, IL

REPORT- **LV-C** DETAILS OF SPACE

DATA FOR SPACE	MODULE	
LOCATION OF ORIGIN IN		
BUILDING COORDINATES		

	AZIMUTH (DEG)	SPACE HEIGHT	MULTIPLIER	AREA (FT)	VOLUME
XB (FT) YB (FT) ZB (FT)					
0.00 0.00 0.00	0.00	9.00	1.0	9.00	225.00

NUMBER OF EXTERIOR SURFACES	NUMBER OF INTERIOR SURFACES	NUMBER OF UNDERGROUND SURFACES	MODULE	
OF SURFACES	6	5	0	

NUMBER OF EXTERIOR SURFACES		
6	1	5

NUMBER OF SUBSURFACES			
EXTERIOR	INTERIOR	DAYLIGHTING	SUNSPACE
		YES	NO
TOTAL WINDOWS	DOORS	WINDOWS	
1	0	0	

TOTAL WINDOWS 1

CALCULATION

FLOOR WEIGHT (LB/SQFT)	TEMPERATURE (F)
70.0	72.0

INFILTRATION

INFILTRATION CALCULATION METHOD   AIR-CHANGE

SCHEDULE   FLOW RATE AIR CHANGES (CFM/SQFT)   0.00
SCH-INFILTR-WNDW   AIR-CHANGE PER HOUR   0.08

HEIGHT TO NEUTRAL ZONE (FT)   0.0

LIGHTING LOAD

LIGHTING (WATTS/)   TYPE   REC-FLUOR-RV   LOAD (KW)   0.00
SCHEDULE   SCH-OFC-LTG
LIGHTING LOAD FRACTION OF LOAD
AREA (SQFT)   SQFT)   1.20   TO SPACE   0.80

INTERIOR SURFACES (U-VALUE INCLUDES BOTH AIR FILMS)

SURFACE	AREA (SQFT)	CONSTRUCTION	U-VALUE (BTU/HR-SQFT-F)	ADJACENT SPACE	SURFACE-TYPE
LEFTPART	135.00	MODULE-PARTITION	0.500	QUICK	ADIABATIC
RIGHTPART	135.00	MODULE-PARTITION	0.500	QUICK	ADIABATIC
BACKPART	135.00	MODULE-PARTITION	0.500	QUICK	ADIABATIC
INTFLR	225.00	MODULE-PARTITION	0.500	QUICK	ADIABATIC
INTCEIL	225.00	MODULE-PARTITION	0.500	QUICK	ADIABATIC

EXTERIOR SURFACES (U-VALUE EXCLUDES OUTSIDE AIR FILM)

SURFACE	AREA (SQFT)	WIDTH (FT)	HEIGHT (FT)	CONSTRUCTION	U-VALUE (BTU/HR-SQFT-F)	SURFACE-TYPE
MODULE-WL	210.00	15.00	14.00	MODULE-WALL	0.080	QUICK

MULTIPLIER   1.0

LOCATION OF ORIGIN IN BUILDING COORDINATES / SPACE COORDINATES

TILT (DEG)   90.0
AZIMUTH (DEG)   0.0

	XB (FT)	YB (FT)	ZB (FT)	X (FT)	Y (FT)	Z (FT)
SURFACE MODULE-WL.	0.00	0.00	0.00	0.00	0.00	0.00

EXTERIOR WINDOWS (U-VALUE INCLUDES OUTSIDE AIR FILM)

WINDOW	AREA (SQFT)	GLASS SHADING COEFF	NUMBER OF PANES	GLASS TYPE CODE	SET-BACK (FT)	GLASS WIDTH (FT)	GLASS HEIGHT (FT)	CENTER-OF-GLASS U-VALUE (BTU/HR-SQFT-F)	GLASS VISIBLE TRANS
MODULE-WIN	146.25	0.95	1	1	0.00	15.00	9.75	0.898	0.800

MULTIPLIER   1.0

LOCATION OF ORIGIN IN SURFACE COORDINATES

WINDOW	SURFACE	XB (FT)	YB (FT)	X (FT)	Y (FT)
MODULE-WIN	MODULE-WL	0.00	0.00	0.00	0.00

REPORT- LV-D DETAILS OF EXTERIOR SURFACES IN THE PROJECT   WEATHER FILE- CHICAGO, IL

NUMBER OF EXTERIOR SURFACES 1   RECTANGULAR 1   OTHER 0
(U-VALUE INCLUDES OUTSIDE AIR FILM; WINDOW INCLUDES FRAME, IF DEFINED)

SURFACE	SPACE	- - - W I N D O W S - - -		- - - - W A L L - - - -		-W A L L + W I N D O W S-		
		U-VALUE	AREA	U-VALUE	AREA	U-VALUE	AREA	AZIMUTH
		(BTU/HR-SQFT-F)	(SQFT)	(BTU/HR-SQFT-F)	(SQFT)	(BTU/HR-SQFT-F)	(SQFT)	
MODULE-WL	MODULE	0.898	146.25	0.078	63.75	0.649	210.00	SOUTH- EAST

	AVERAGE U-VALUE/WINDOWS	AVERAGE U-VALUE/WALLS	AVERAGE U-VALUE WALLS+WINDOWS	WINDOW AREA	WALL AREA	WINDOW+WALL AREA
	(BTU/HR-SQFT-F)	(BTU/HR-SQFT-F)	(BTU/HR-SQFT-F)	(SQFT)	(SQFT)	(SQFT)
SOUTH-EAST	0.898	0.078	0.649	146.25	63.75	210.00
ALL WALLS	0.898	0.078	0.649	146.25	63.75	210.00
WALLS+ROOFS	0.898	0.078	0.649	146.25	63.75	210.00
BUILDING	0.898	0.078	0.649	146.25	63.75	210.00

REPORT- LV-F DETAILS OF INTERIOR SURFACES IN THE PROJECT   WEATHER FILE- CHICAGO, IL

NUMBER OF INTERIOR SURFACES 5
(U-VALUE INCLUDES BOTH AIR FILMS)

SURFACE NAME	AREA (SQFT)	CONSTRUCTION NAME	SURFACE TYPE	U-VALUE (BTU/HR-SQFT-F)	ADJACENT SPACES	
					SPACE-1	SPACE-2
LEFTPART	135.00	MODULE-PARTITION QUICK	ADIABATIC	0.500	MODULE	
RIGHTPART	135.00	MODULE-PARTITION QUICK	ADIABATIC	0.500	MODULE	
BACKPART	135.00	MODULE-PARTITION QUICK	ADIABATIC	0.500	MODULE	
INTFLR	225.00	MODULE-PARTITION QUICK	ADIABATIC	0.500	MODULE	
INTCEIL	225.00	MODULE-PARTITION QUICK	ADIABATIC	0.500	MODULE	

REPORT- LV-G   DETAILS OF SCHEDULES OCCURRING IN THE PROJECT   WEATHER FILE- CHICAGO, IL.

(NON DIMENSIONLESS SCHEDULES ARE GIVEN IN ENGLISH UNITS)

NUMBER OF SCHEDULES 3

SCHEDULE SCH-OFC-LTG
THROUGH 31 12

FOR DAYS SUN

HOUR	1	2	3	4	5	6	7	8	9	10	11	12	13	14	15	16	17	18	19	20	21	22	23	24
	0.05	0.05	0.05	0.05	0.05	0.05	0.05	0.05	0.05	0.05	0.05	0.05	0.05	0.05	0.05	0.05	0.05	0.05	0.05	0.05	0.05	0.05	0.05	0.05

FOR DAYS MON TUE WED THU FRI

HOUR	1	2	3	4	5	6	7	8	9	10	11	12	13	14	15	16	17	18	19	20	21	22	23	24
	0.05	0.05	0.05	0.05	0.05	0.10	0.35	0.50	0.90	0.90	0.90	0.90	0.90	0.90	0.90	0.90	0.90	0.90	0.90	0.50	0.35	0.35	0.10	0.05

FOR DAYS SAT

HOUR	1	2	3	4	5	6	7	8	9	10	11	12	13	14	15	16	17	18	19	20	21	22	23	24
	0.05	0.05	0.05	0.05	0.05	0.05	0.05	0.05	0.05	0.05	0.05	0.05	0.05	0.05	0.05	0.05	0.05	0.05	0.05	0.05	0.05	0.05	0.05	0.05

FOR DAYS HOL

HOUR	1	2	3	4	5	6	7	8	9	10	11	12	13	14	15	16	17	18	19	20	21	22	23	24
	0.05	0.05	0.05	0.05	0.05	0.05	0.05	0.05	0.05	0.05	0.05	0.05	0.05	0.05	0.05	0.05	0.05	0.05	0.05	0.05	0.05	0.05	0.05	0.05

SCHEDULE SCH-INFLTR-WNDW

THROUGH 28 2

FOR DAYS SUN MON TUE WED THU FRI SAT HOL

HOUR	1	2	3	4	5	6	7	8	9	10	11	12	13	14	15	16	17	18	19	20	21	22	23	24
	1.00	1.00	1.00	1.00	1.00	1.00	1.00	1.00	1.00	1.00	1.00	1.00	1.00	1.00	1.00	1.00	1.00	1.00	1.00	1.00	1.00	1.00	1.00	1.00

THROUGH 30 4

FOR DAYS SUN MON TUE WED THU FRI SAT HOL

HOUR	1	2	3	4	5	6	7	8	9	10	11	12	13	14	15	16	17	18	19	20	21	22	23	24
	0.50	0.50	0.50	0.50	0.50	0.50	0.50	0.50	0.50	0.50	0.50	0.50	0.50	0.50	0.50	0.50	0.50	0.50	0.50	0.50	0.50	0.50	0.50	0.50

THROUGH 31 10

FOR DAYS SUN MON TUE WED THU FRI SAT HOL

HOUR	1	2	3	4	5	6	7	8	9	10	11	12	13	14	15	16	17	18	19	20	21	22	23	24
	0.00	0.00	0.00	0.00	0.00	0.00	0.00	0.00	0.00	0.00	0.00	0.00	0.00	0.00	0.00	0.00	0.00	0.00	0.00	0.00	0.00	0.00	0.00	0.00

THROUGH 31 12

FOR DAYS SUN MON TUE WED THU FRI SAT HOL

HOUR	1	2	3	4	5	6	7	8	9	10	11	12	13	14	15	16	17	18	19	20	21	22	23	24
	0.50	0.50	0.50	0.50	0.50	0.50	0.50	0.50	0.50	0.50	0.50	0.50	0.50	0.50	0.50	0.50	0.50	0.50	0.50	0.50	0.50	0.50	0.50	0.50

SCHEDULE SCH-DAYREP-ON
THROUGH 31 12

FOR DAYS SUN MON TUE WED THU FRI SAT HOL

HOUR	1	2	3	4	5	6	7	8	9	10	11	12	13	14	15	16	17	18	19	20	21	22	23	24
	1.00	1.00	1.00	1.00	1.00	1.00	1.00	1.00	1.00	1.00	1.00	1.00	1.00	1.00	1.00	1.00	1.00	1.00	1.00	1.00	1.00	1.00	1.00	1.00
	0.00	0.00	0.00	0.00	1.00	1.00	1.00	1.00	1.00	1.00	1.00	1.00	1.00	1.00	1.00	1.00	1.00	1.00	1.00	1.00	0.00	0.00	0.00	0.00

214

NUMBER OF WINDOWS 1   RECTANGULAR 1   OTHER 0

RECTANGULAR WINDOWS (U-VALUES INCLUDE OUTSIDE AIR FILM)

WINDOW NAME	MULTIPLIER	GLASS AREA (SQFT)	GLASS HEIGHT (FT)	GLASS WIDTH (FT)	LOCATION OF ORIGIN IN SURFACE COORDINATES X (FT) Y (FT)		FRAME AREA (SQFT)	FRAME U-VALUE (BTU/HR-SQFT-F)
MODULE-WIN	1.0	146.25	9.75	15.00	0.00	0.00	0.00	0.400

WINDOW NAME	SETBACK (FT)	X-DIVISIONS	GLASS SHADING COEFF	GLASS NUMBER OF PANES	GLASS TYPE CODE	INFILTRATION FLOW COEFF	GLASS U-VALUE (BTU/HR-SQFT-F)	CENTER-OF-GLASS VISIBLE TRANS
MODULE-WIN	0.00	10	0.95	1	1	0.0	0.898	0.800

REPORT- IV-I DETAILS OF CONSTRUCTIONS OCCURRING IN THE PROJECT   WEATHER FILE- CHICAGO, IL

NUMBER OF CONSTRUCTIONS 2   DELAYED 0   QUICK 2

CONSTRUCTION NAME	U-VALUE (BTU/HR-SQFT-F)	SURFACE ABSORPTANCE	SURFACE ROUGHNESS INDEX	SURFACE TYPE	NUMBER OF RESPONSE FACTORS
MODULE-WALL	0.080	0.70	3	QUICK	0
MODULE-PARTITION	0.500	0.70	3	QUICK	0

APPENDIX D

EXAMPLE OF DOE2.1E SIMULATION FILE-SIM FILE

216

SPACE NAME	MULTIPLIER	COOLING LOAD	TIME OF	DRY-	WET-	HEATING LOAD	TIME OF	DRY-	WET-
	SPACE FLOOR	(KBTU/HR)	PEAK	BULB	BULB	(KBTU/HR)	PEAK	BULB	BULB
MODULE	1. 1.	7.945	OCT 2 12 NOON	71.F	54.F	-8.688	JAN 27 4 AM	-7.F	-8.F

SUM

| | 7.945 | | | | -8.688 | | | |

BUILDING PEAK

| | 7.945 | OCT 2 12 NOON | 71.F | 54.F | -8.688 | JAN 27 4 AM | -7.F | -8.F |

REPORT- LS-B SPACE PEAK LOAD COMPONENTS   WEATHER FILE- CHICAGO, IL

SPACE   MODULE

SPACE TEMPERATURE USED FOR THE LOADS CALCULATION IS   72 F /   22 C

	MULTIPLIER	FLOOR MULTIPLIER		
	FLOOR AREA	225 SQFT	21 M2	1.0
	VOLUME	2025 CUFT	57 M3	

	COOLING LOAD		HEATING LOAD	
TIME	OCT 2 12NOON		JAN 27 4AM	
DRY-BULB TEMP	71 F	22 C	-7 F	-22 C
WET-BULB TEMP	54 F	12 C	-8 F	-22 C
TOT HORIZONTAL SOLAR RAD	210 BTU/H.SQFT	662 W/M2	0 BTU/H.SQFT	0 W/M2
WINDSPEED AT SPACE	7.9 KTS	4.0 M/S	8.5 KTS	4.4 M/S
CLOUD AMOUNT 0(CLEAR)-10	0		0	

	SENSIBLE		LATENT		SENSIBLE	
	(KBTU/H)	(KW)	(KBTU/H)	(KW)	(KBTU/H)	(KW)
WALL CONDUCTION	0.153	0.045	0.000	0.000	-0.384	-0.112
ROOF CONDUCTION	0.000	0.000	0.000	0.000	0.000	0.000
WINDOW GLASS+FRM COND	-0.350	-0.103	0.000	0.000	-8.242	-2.415
WINDOW GLASS SOLAR	7.958	2.332	0.000	0.000	0.119	0.035
DOOR CONDUCTION	0.000	0.000	0.000	0.000	0.000	0.000
INTERNAL SURFACE COND	0.000	0.000	0.000	0.000	0.000	0.000
UNDERGROUND SURF COND	0.000	0.000	0.000	0.000	0.000	0.000
OCCUPANTS TO SPACE	0.000	0.000	0.000	0.000	0.000	0.000
LIGHT TO SPACE	0.184	0.054	0.000	0.000	0.036	0.011
EQUIPMENT TO SPACE	0.000	0.000	0.000	0.000	0.000	0.000
PROCESS TO SPACE	0.000	0.000	0.000	0.000	0.000	0.000
INFILTRATION	0.000	0.000	0.000	0.000	-0.217	-0.064
TOTAL	7.945	2.328	0.000	0.000	-8.688	-2.546
TOTAL / AREA	0.035	0.111	0.000	0.000	-0.039	-0.122
TOTAL LOAD	7.945 KBTU/H	2.328 KW			-8.688 KBTU/H	-2.546 KW
TOTAL LOAD / AREA	35.31 BTU/H.SQFT	111.372 W/M2			38.612 BTU/H.SQFT	121.777 W/M2

REPORT- LS-D BUILDING MONTHLY LOADS SUMMARY

WEATHER FILE- CHICAGO, IL

MONTH	--------COOLING--------					--------HEATING--------					----ELEC----	
	COOLING ENERGY (MBTU)	TIME OF MAX DY HR	DRY-BULB TEMP	WET-BULB TEMP	MAXIMUM COOLING LOAD (KBTU/HR)	HEATING ENERGY (MBTU)	TIME OF MAX DY HR	DRY-BULB TEMP	WET-BULB TEMP	MAXIMUM HEATING LOAD (KBTU/HR)	ELEC-TRICAL ENERGY (KWH)	MAXIMUM ELEC LOAD (KW)
JAN	0.14101	8 12	43.F	37.F	4.574	-2.535	27 4	-7.F	-8.F	-8.688	52.	0.243
FEB	0.30290	22 12	45.F	37.F	6.847	-1.951	19 7	1.F	0.F	-7.164	42.	0.243
MAR	0.53447	18 11	58.F	45.F	7.904	-1.469	4 6	12.F	9.F	-5.949	42.	0.243
APR	0.72261	16 11	69.F	53.F	7.441	-0.816	11 3	28.F	26.F	-4.188	37.	0.243
MAY	1.09782	12 11	86.F	73.F	6.876	-0.292	16 5	39.F	37.F	-2.921	34.	0.205
JUN	1.55609	9 13	94.F	72.F	6.166	-0.061	28 4	54.F	52.F	-1.596	32.	0.198
JUL	1.85843	1 15	99.F	75.F	6.421	-0.013	10 5	55.F	47.F	-0.820	32.	0.161
AUG	1.80641	30 14	94.F	75.F	6.398	-0.024	25 5	51.F	47.F	-1.554	35.	0.219
SEP	1.48877	15 10	79.F	70.F	7.576	-0.138	28 6	45.F	43.F	-1.979	38.	0.241
OCT	1.00413	2 11	71.F	54.F	7.945	-0.438	29 6	37.F	36.F	-3.170	46.	0.243
NOV	0.39448	3 10	65.F	54.F	6.829	-1.327	29 5	20.F	18.F	-5.647	45.	0.243
DEC	0.07432	18 11	41.F	37.F	3.332	-2.580	2 7	5.F	3.F	-6.978	57.	0.243
TOTAL	10.981					-11.644					491.	
MAX					7.945					-8.688		0.243

218

REPORT- LS-E   SPACE MONTHLY LOAD COMPONENTS IN MBTU FOR   MODULE   WEATHER FILE- CHICAGO, IL

(UNITS=MBTU)		WALLS	ROOFS	INT SUR	UND SUR	INFIL	WIN CON	WIN SOL	OCCUP	LIGHTS	EQUIP	SOURCE	TOTAL
JAN	HEATNG	-0.147	0.000	0.000	0.000	-0.070	-3.273	0.832	0.000	0.122	0.000	0.000	-2.535
	SEN CL	-0.011	0.000	0.000	0.000	-0.009	-0.410	0.553	0.000	0.018	0.000	0.000	0.141
	LAT CL		0.000	0.000	0.000	0.000			0.000		0.000	0.000	0.000
FEB	HEATNG	-0.119	0.000	0.000	0.000	-0.054	-2.675	0.807	0.000	0.090	0.000	0.000	-1.951
	SEN CL	-0.014	0.000	0.000	0.000	-0.012	-0.573	0.878	0.000	0.023	0.000	0.000	0.303
	LAT CL		0.000	0.000	0.000	0.000			0.000		0.000	0.000	0.000
MAR	HEATNG	-0.100	0.000	0.000	0.000	-0.023	-2.256	0.823	0.000	0.086	0.000	0.000	-1.469
	SEN CL	-0.014	0.000	0.000	0.000	-0.007	-0.674	1.200	0.000	0.029	0.000	0.000	0.534
	LAT CL		0.000	0.000	0.000	0.000			0.000		0.000	0.000	0.000
APR	HEATNG	-0.057	0.000	0.000	0.000	-0.012	-1.306	0.505	0.000	0.054	0.000	0.000	-0.816
	SEN CL	-0.006	0.000	0.000	0.000	-0.004	-0.472	1.156	0.000	0.048	0.000	0.000	0.723
	LAT CL		0.000	0.000	0.000	0.000			0.000		0.000	0.000	0.000
MAY	HEATNG	-0.026	0.000	0.000	0.000	0.000	-0.568	0.274	0.000	0.027	0.000	0.000	-0.292
	SEN CL	0.004	0.000	0.000	0.000	0.000	-0.364	1.392	0.000	0.066	0.000	0.000	1.098
	LAT CL		0.000	0.000	0.000	0.000			0.000		0.000	0.000	0.000
JUN	HEATNG	-0.007	0.000	0.000	0.000	0.000	-0.147	0.081	0.000	0.011	0.000	0.000	-0.061
	SEN CL	0.025	0.000	0.000	0.000	0.000	0.096	1.358	0.000	0.076	0.000	0.000	1.556
	LAT CL		0.000	0.000	0.000	0.000			0.000		0.000	0.000	0.000
JUL	HEATNG	-0.002	0.000	0.000	0.000	0.000	-0.041	0.027	0.000	0.003	0.000	0.000	-0.013
	SEN CL	0.036	0.000	0.000	0.000	0.000	0.231	1.506	0.000	0.085	0.000	0.000	1.858
	LAT CL		0.000	0.000	0.000	0.000			0.000		0.000	0.000	0.000
AUG	HEATNG	-0.003	0.000	0.000	0.000	0.000	-0.065	0.040	0.000	0.005	0.000	0.000	-0.024
	SEN CL	0.034	0.000	0.000	0.000	0.000	0.159	1.523	0.000	0.091	0.000	0.000	1.806
	LAT CL		0.000	0.000	0.000	0.000			0.000		0.000	0.000	0.000
SEP	HEATNG	-0.013	0.000	0.000	0.000	0.000	-0.296	0.148	0.000	0.023	0.000	0.000	-0.138
	SEN CL	0.017	0.000	0.000	0.000	0.000	-0.150	1.540	0.000	0.081	0.000	0.000	1.489
	LAT CL		0.000	0.000	0.000	0.000			0.000		0.000	0.000	0.000
OCT	HEATNG	-0.035	0.000	0.000	0.000	0.000	-0.755	0.295	0.000	0.058	0.000	0.000	-0.438
	SEN CL	-0.002	0.000	0.000	0.000	0.000	-0.505	1.443	0.000	0.068	0.000	0.000	1.004
	LAT CL		0.000	0.000	0.000	0.000			0.000		0.000	0.000	0.000

NOV	HEATING	-0.084	0.000	0.000	-0.018	-1.858	0.541	0.000	0.092	0.000	-1.327
	SEN CL	-0.008	0.000	0.000	-0.004	-0.415	0.790	0.000	0.000	0.000	0.394
	LAT CL		0.000	0.000	0.000			0.000	0.031	0.000	0.000
DEC	HEATING	-0.145	0.000	0.000	-0.060	-3.173	0.653	0.000	0.145	0.000	-2.580
	SEN CL	-0.006	0.000	0.000	-0.005	-0.239	0.315	0.000	0.009	0.000	0.074
	LAT CL		0.000	0.000	0.000			0.000	0.031	0.000	0.000
TOT	HEATING	-0.737	0.000	0.000	-0.238	-16.411	5.025	0.000	0.716	0.000	-11.644
	SEN CL	0.056	0.000	0.000	-0.040	-3.314	13.655	0.000	0.625	0.000	10.981
	LAT CL	**0.000**	**0.000**	**0.000**	**0.000**			0.000	0.000	0.000	0.000

REPORT- LS-G SPACE DAYLIGHTING SUMMARY   WEATHER FILE- CHICAGO, IL

SPACE MODULE

------REPORT SCHEDULE HOURS WITH SUN UP------

MONTH	PERCENT LIGHTING ENERGY REDUCTION BY DAYLIGHTING (ALL HOURS)			PERCENT LIGHTING ENERGY REDUCTION BY DAYLIGHTING (REPORT SCHEDULE HOURS)			AVERAGE DAYLIGHT ILLUMINANCE (FOOTCANDLES)		PERCENT HOURS DAYLIGHT ILLUMINANCE ABOVE SETPOINT		AVERAGE GLARE INDEX		PERCENT HOURS GLARE TOO HIGH	
	TOTAL ZONE	REF PT 1	REF PT 2	TOTAL ZONE	REF PT 1	REF PT 2	REF PT 1	REF PT 2	REF PT 1	REF PT 2	REF PT 1	REF PT 2	REF PT 1	REF PT 2
JAN	36.3	36.3	0.0	45.4	45.4	0.0	93.3	0.0	39.4	0.0	0.0	0.0	0.0	0.0
FEB	41.3	41.3	0.0	51.6	51.6	0.0	102.8	0.0	49.8	0.0	0.0	0.0	0.0	0.0
MAR	46.9	46.9	0.0	58.5	58.5	0.0	107.5	0.0	56.1	0.0	0.0	0.0	0.0	0.0
APR	54.0	54.0	0.0	64.1	64.1	0.0	97.3	0.0	67.1	0.0	0.0	0.0	0.0	0.0
MAY	57.4	57.4	0.0	67.0	67.0	0.0	99.3	0.0	68.5	0.0	0.0	0.0	0.0	0.0
JUN	58.4	58.4	0.0	66.9	66.9	0.0	99.1	0.0	74.6	0.0	0.0	0.0	0.0	0.0
JUL	60.1	60.1	0.0	68.3	68.3	0.0	105.4	0.0	80.1	0.0	0.0	0.0	0.0	0.0
AUG	55.9	55.9	0.0	65.5	65.5	0.0	111.2	0.0	80.1	0.0	0.0	0.0	0.0	0.0
SEP	50.6	50.6	0.0	62.0	62.0	0.0	113.1	0.0	69.4	0.0	0.0	0.0	0.0	0.0
OCT	43.2	43.2	0.0	54.0	54.0	0.0	115.4	0.0	54.5	0.0	0.0	0.0	0.0	0.0
NOV	35.5	35.5	0.0	44.4	44.4	0.0	91.4	0.0	39.1	0.0	0.0	0.0	0.0	0.0
DEC	29.8	29.8	0.0	37.3	37.3	0.0	70.9	0.0	28.1	0.0	0.0	0.0	0.0	0.0
ANNUAL	47.6	47.6	0.0	57.2	57.2	0.0	101.1	0.0	60.3	0.0	0.0	0.0	0.0	0.0

REPORT- LS-H PERCENT LIGHTING ENERGY REDUCTION BY DAYLIGHT MODULE

WEATHER FILE- CHICAGO, IL

SPACE MODULE

MONTH	1	2	3	4	5	6	7	8	9	10	11	12	13	14	15	16	17	18	19	20	21	22	23	24	ALL HOURS
JAN	0	0	0	0	0	0	0	14	54	62	66	66	64	62	58	46	8	0	0	0	0	0	0	0	36
FEB	0	0	0	0	0	0	0	44	54	64	67	67	67	65	65	56	35	2	0	0	0	0	0	0	41
MAR	0	0	0	0	0	0	27	54	68	68	69	68	67	67	60	52	16	0	0	0	0	0	0	0	47
APR	0	0	0	0	0	2	15	53	63	70	70	70	70	70	68	58	42	1	0	0	0	0	0	0	54
MAY	0	0	0	0	0	6	46	65	69	69	70	70	70	70	69	68	65	48	11	0	0	0	0	0	57
JUN	0	0	0	0	2	54	62	67	68	69	70	70	70	70	69	68	63	57	26	0	0	0	0	0	58
JUL	0	0	0	0	2	54	62	67	68	70	70	70	70	70	69	67	63	57	26	0	0	0	0	0	60
AUG	0	0	0	0	0	48	69	67	68	69	70	70	70	69	69	64	50	14	2	0	0	0	0	0	56
SEP	0	0	0	0	0	13	51	61	67	67	68	70	69	69	64	61	52	0	0	0	0	0	0	0	51
OCT	0	0	0	0	0	3	50	56	63	66	65	68	68	66	61	52	22	0	0	0	0	0	0	0	43
NOV	0	0	0	0	0	0	27	46	55	62	65	65	65	61	53	28	2	0	0	0	0	0	0	0	35
DEC	0	0	0	0	0	0	2	12	39	51	59	63	58	54	49	26	0	0	0	0	0	0	0	0	30
ANNUAL	0	0	0	0	1	21	38	56	62	66	67	68	67	66	64	56	41	25	4	0	0	0	0	0	48

NOTE- THE ENTRIES IN THIS REPORT ARE NOT SUBJECT TO THE DAYLIGHTING REPORT SCHEDULE

REPORT- LS-J DAYLIGHT ILLUMINANCE FREQUENCY OF OCCURENCE MODULE

WEATHER FILE- CHICAGO, IL

SPACE
MODULE

PERCENT OF HOURS IN ILLUMINANCE RANGE — ILLUMINANCE RANGE (FOOTCANDLES)

MONTH	REF PT	0--10	10--20	20--30	30--40	40--50	50--60	60--70	70--80	80-ABOVE
JAN	-1-	18	20	8	9	5	2	3	1	34
	-2-	0	0	0	0	0	0	0	0	0
FEB	-1-	21	7	11	5	6	10	2	1	36
	-2-	0	0	0	0	0	0	0	0	0
MAR	-1-	11	8	9	9	7	6	9	3	38
	-2-	0	0	0	0	0	0	0	0	0
APR	-1-	8	6	7	7	6	5	8	8	47
	-2-	0	0	0	0	0	0	0	0	0
MAY	-1-	10	0	4	9	8	4	5	7	52
	-2-	0	0	0	0	0	0	0	0	0
JUN	-1-	9	3	4	4	5	1	3	8	59
	-2-	0	0	0	0	0	0	0	0	0
JUL	-1-	9	2	3	5	1	1	4	8	67
	-2-	0	0	0	0	0	0	0	0	0
AUG	-1-	8	6	3	5	2	5	7	6	63
	-2-	0	0	0	0	0	0	0	0	0
SEP	-1-	9	5	5	5	6	5	7	5	53
	-2-	0	0	0	0	0	0	0	0	0
OCT	-1-	11	10	7	9	7	4	3	3	45
	-2-	0	0	0	0	0	0	0	0	0
NOV	-1-	22	12	7	11	9	1	2	4	32
	-2-	0	0	0	0	0	0	0	0	0
DEC	-1-	27	19	13	12	1	2	1	2	24
	-2-	0	0	0	0	0	0	0	0	0
ANNUAL	-1-	13	8	7	7	5	4	5	5	47
	-2-	0	0	0	0	0	0	0	0	0

PERCENT OF HOURS ILLUMINANCE LEVEL EXCEEDED — ILLUMINANCE LEVEL (FOOTCANDLES)

MONTH	REF PT	0	10	20	30	40	50	60	70	80
JAN	-1-	100	82	62	54	45	39	37	35	34
	-2-	0	0	0	0	0	0	0	0	0
FEB	-1-	100	79	72	61	56	50	39	37	36
	-2-	0	0	0	0	0	0	0	0	0
MAR	-1-	100	89	81	72	63	56	50	42	38
	-2-	0	0	0	0	0	0	0	0	0
APR	-1-	100	92	86	80	73	67	62	54	47
	-2-	0	0	0	0	0	0	0	0	0
MAY	-1-	100	90	89	86	76	68	64	59	52
	-2-	0	0	0	0	0	0	0	0	0
JUN	-1-	100	91	88	84	80	75	71	67	59
	-2-	0	0	0	0	0	0	0	0	0
JUL	-1-	100	91	89	87	82	80	79	75	67
	-2-	0	0	0	0	0	0	0	0	0
AUG	-1-	100	92	87	85	82	80	75	69	63
	-2-	0	0	0	0	0	0	0	0	0
SEP	-1-	100	91	86	81	76	69	65	57	53
	-2-	0	0	0	0	0	0	0	0	0
OCT	-1-	100	89	79	70	61	54	50	48	45
	-2-	0	0	0	0	0	0	0	0	0
NOV	-1-	100	78	66	59	48	39	38	37	32
	-2-	0	0	0	0	0	0	0	0	0
DEC	-1-	100	73	55	41	29	28	26	26	24
	-2-	0	0	0	0	0	0	0	0	0
ANNUAL	-1-	100	87	79	73	66	60	56	52	47
	-2-	0	0	0	0	0	0	0	0	0

NOTE- THE HOURS CONSIDERED IN THIS REPORT ARE THOSE
WITH SUN UP AND DAYLIGHTING REPORT SCHEDULE ON

REPORT- LS-D BUILDING MONTHLY LOADS SUMMARY WEATHER FILE- TRY BANGKOK

MONTH	COOLING ENERGY (MBTU)	TIME OF MAX DY	HR	DRY-BULB TEMP	WET-BULB TEMP	MAXIMUM COOLING LOAD (KBTU/HR)	HEATING ENERGY (MBTU)	TIME OF MAX DY	HR	DRY-BULB TEMP	WET-BULB TEMP	MAXIMUM HEATING LOAD (KBTU/HR)	ELECTRICAL ENERGY (KWH)	MAXIMUM ELEC LOAD (KW)
						— COOLING —						— HEATING —		— ELEC —
JAN	0.54918	10	15	95.F	80.F	1.667	0.000					-0.097	67.	0.243
FEB	0.52573	23	15	97.F	83.F	1.838	-0.001	28	4	68.F	64.F	-0.196	56.	0.243
MAR	0.70088	29	15	95.F	78.F	1.974	0.000					0.000	63.	0.243
APR	0.74285	9	15	100.F	81.F	2.122	0.000					0.000	58.	0.226
MAY	0.82044	30	15	95.F	82.F	2.212	0.000					0.000	59.	0.225
JUN	0.77609	7	15	93.F	82.F	2.187	0.000					0.000	58.	0.223
JUL	0.78789	30	16	95.F	80.F	2.103	0.000					0.000	57.	0.228
AUG	0.74797	2	16	99.F	82.F	2.164	0.000					0.000	62.	0.226
SEP	0.63591	21	15	88.F	81.F	1.844	0.000					0.000	55.	0.234
OCT	0.62744	4	14	92.F	81.F	1.776	0.000					0.000	63.	0.243
NOV	0.48027	29	14	93.F	80.F	1.585	-0.001	26	5	68.F	65.F	-0.103	62.	0.243
DEC	0.51359	12	14	90.F	77.F	1.567	-0.001	26	6	66.F	64.F	-0.158	64.	0.243
TOTAL	7.908						-0.003						725.	
MAX						2.212						-0.196		0.243

REPORT- LS-F BUILDING MONTHLY LOAD COMPONENTS IN MBTU WEATHER FILE- TRY BANGKOK

(UNITS=MBTU)	WALLS	ROOFS	INT SUR	UND SUR	INFIL	WIN CON	WIN SOL	OCCUP	LIGHTS	EQUIP	SOURCE	TOTAL
JAN HEATING	-0.001	0.000	0.000	0.000	0.000	-0.001	0.000	0.000	0.000	0.000	0.000	0.000
JAN SEN CL	0.107	0.000	0.000	0.000	0.005	0.053	0.202	0.182	0.000	0.000	0.000	0.549
JAN LAT CL	0.000	0.000	0.000	0.000	0.027	0.000	0.000	0.000	0.000	0.000	0.000	0.027
FEB HEATING	-0.002	0.000	0.000	0.000	0.000	-0.002	0.001	0.002	0.000	0.000	0.000	-0.001
FEB SEN CL	0.108	0.000	0.000	0.000	0.004	0.050	0.211	0.152	0.000	0.000	0.000	0.526
FEB LAT CL	0.000	0.000	0.000	0.000	0.019	0.000	0.000	0.000	0.000	0.000	0.000	0.019
MAR HEATING	0.000	0.000	0.000	0.000	0.000	0.000	0.000	0.000	0.000	0.000	0.000	0.000
MAR SEN CL	0.162	0.000	0.000	0.000	0.003	0.092	0.272	0.172	0.000	0.000	0.000	0.701
MAR LAT CL	0.000	0.000	0.000	0.000	0.013	0.000	0.000	0.000	0.000	0.000	0.000	0.013

DOE-2.1E-136 10/05/2004 14:06:09 LDL RUN 1

APR	HEATNG	0.000	0.000	0.000	0.000	0.000	0.000	0.000	0.000	0.000	0.000	-0.001
	SEN CL	0.174	0.000	0.000	0.004	0.095	0.313	0.000	0.157	0.000	0.000	0.743
	LAT CL	0.000	0.000	0.000	0.019	0.000	0.000	0.000	0.000	0.000	0.000	0.019
MAY	HEATNG	0.000	0.000	0.000	0.000	0.000	0.000	0.000	0.000	0.000	0.000	0.000
	SEN CL	0.188	0.000	0.000	0.000	0.101	0.371	0.000	0.161	0.000	0.000	0.820
	LAT CL	0.000	0.000	0.000	0.019	0.000	0.000	0.000	0.000	0.000	0.000	0.019
JUN	HEATNG	0.000	0.000	0.000	0.000	0.000	0.000	0.000	0.000	0.000	0.000	0.000
	SEN CL	0.160	0.000	0.000	0.000	0.081	0.375	0.000	0.160	0.000	0.000	0.776
	LAT CL	0.000	0.000	0.000	0.000	0.000	0.000	0.000	0.000	0.000	0.000	0.000
JUL	HEATNG	0.000	0.000	0.000	0.000	0.000	0.000	0.000	0.000	0.000	0.000	0.000
	SEN CL	0.165	0.000	0.000	0.000	0.085	0.381	0.000	0.156	0.000	0.000	0.788
	LAT CL	0.000	0.000	0.000	0.000	0.000	0.000	0.000	0.000	0.000	0.000	0.000
AUG	HEATNG	0.000	0.000	0.000	0.000	0.000	0.000	0.000	0.000	0.000	0.000	0.000
	SEN CL	0.162	0.000	0.000	0.000	0.083	0.333	0.000	0.170	0.000	0.000	0.748
	LAT CL	0.000	0.000	0.000	0.000	0.000	0.000	0.000	0.000	0.000	0.000	0.000
SEP	HEATNG	0.000	0.000	0.000	0.000	0.000	0.000	0.000	0.000	0.000	0.000	0.000
	SEN CL	0.136	0.000	0.000	0.000	0.064	0.285	0.000	0.151	0.000	0.000	0.636
	LAT CL	0.000	0.000	0.000	0.000	0.000	0.000	0.000	0.000	0.000	0.000	0.000
OCT	HEATNG	0.000	0.000	0.000	0.000	0.000	0.000	0.000	0.000	0.000	0.000	-0.001
	SEN CL	0.136	0.000	0.000	0.000	0.063	0.257	0.000	0.172	0.000	0.000	0.627
	LAT CL	0.000	0.000	0.000	0.000	0.000	0.000	0.000	0.000	0.000	0.000	0.000
NOV	HEATNG	-0.001	0.000	0.000	0.000	0.000	0.000	0.000	0.000	0.000	0.000	-0.001
	SEN CL	0.088	0.000	0.000	0.001	0.028	0.196	0.000	0.167	0.000	0.000	0.480
	LAT CL	0.000	0.000	0.000	0.007	0.000	0.000	0.000	0.000	0.000	0.000	0.007
DEC	HEATNG	-0.001	0.000	0.000	0.000	0.000	0.000	0.000	0.000	0.000	0.000	-0.001
	SEN CL	0.103	0.000	0.000	0.004	0.046	0.187	0.000	0.174	0.000	0.000	0.514
	LAT CL	0.000	0.000	0.000	0.017	0.000	0.000	0.000	0.000	0.000	0.000	0.017
TOT	HEATNG	-0.006	0.000	0.000	0.000	0.000	0.004	0.000	0.000	0.000	0.000	-0.003
	SEN CL	1.688	0.000	0.000	0.022	0.841	3.383	0.000	1.974	0.000	0.000	7.908
	LAT CL	0.000	0.000	0.000	0.101	0.000	0.000	0.000	0.000	0.000	0.000	0.101

REPORT- LS-G   SPACE DAYLIGHTING SUMMARY   WEATHER FILE- TRY BANGKOK

SPACE MODULE

MONTH	PERCENT LIGHTING ENERGY REDUCTION BY DAYLIGHTING (ALL HOURS)			PERCENT LIGHTING ENERGY REDUCTION BY DAYLIGHTING (REPORT SCHEDULE HOURS)			AVERAGE DAYLIGHT ILLUMINANCE (FOOTCANDLES)		PERCENT HOURS DAYLIGHT ILLUMINANCE ABOVE SETPOINT		AVERAGE GLARE INDEX		PERCENT HOURS GLARE TOO HIGH	
	TOTAL ZONE	REF PT 1	REF PT 2	TOTAL ZONE	REF PT 1	REF PT 2	REF PT 1	REF PT 2	REF PT 1	REF PT 2	REF PT 1	REF PT 2	REF PT 1	REF PT 2
JAN	17.1	17.1	0.0	21.4	21.4	0.0	11.2	0.0	0.0	0.0	0.0	0.0	0.0	0.0
FEB	20.7	20.7	0.0	25.9	25.9	0.0	13.6	0.0	0.0	0.0	0.0	0.0	0.0	0.0
MAR	23.6	23.6	0.0	29.4	29.4	0.0	15.5	0.0	0.0	0.0	0.0	0.0	0.0	0.0
APR	25.6	25.6	0.0	31.3	31.3	0.0	17.8	0.0	0.0	0.0	0.0	0.0	0.0	0.0
MAY	27.1	27.1	0.0	33.1	33.1	0.0	17.3	0.0	0.0	0.0	0.0	0.0	0.0	0.0
JUN	25.7	25.7	0.0	31.1	31.1	0.0	16.3	0.0	0.0	0.0	0.0	0.0	0.0	0.0
JUL	26.3	26.3	0.0	31.9	31.9	0.0	17.6	0.0	0.0	0.0	0.0	0.0	0.0	0.0
AUG	25.9	25.9	0.0	31.6	31.6	0.0	17.9	0.0	0.0	0.0	0.0	0.0	0.0	0.0
SEP	24.1	24.1	0.0	29.7	29.7	0.0	16.9	0.0	0.0	0.0	0.0	0.0	0.0	0.0
OCT	21.7	21.7	0.0	27.0	27.0	0.0	15.3	0.0	0.0	0.0	0.0	0.0	0.0	0.0
NOV	17.5	17.5	0.0	21.8	21.8	0.0	11.5	0.0	0.0	0.0	0.0	0.0	0.0	0.0
DEC	15.6	15.6	0.0	19.5	19.5	0.0	10.3	0.0	0.0	0.0	0.0	0.0	0.0	0.0
ANNUAL	22.6	22.6	0.0	27.9	27.9	0.0	15.0	0.0	0.0	0.0	0.0	0.0	0.0	0.0

------REPORT SCHEDULE HOURS WITH SUN UP------

226

Heating and Cooling

Label				
1/ 1/ 1/ 1/ 1/ 1/ 1/01/C&HE/	0.00000	0.000	-0.904	-2.540
1/ 1/ 1/ 1/ 1/ 1/ 1/02/C&HE/	0.00034	0.094	-0.766	-2.360
1/ 1/ 1/ 1/ 1/ 1/ 1/03/C&HE/	0.00407	0.349	-0.578	-1.979
1/ 1/ 1/ 1/ 1/ 1/ 1/04/C&HE/	0.06818	0.978	-0.264	-1.229
1/ 1/ 1/ 1/ 1/ 1/ 1/05/C&HE/	0.21670	1.466	-0.103	-0.864
1/ 1/ 1/ 1/ 1/ 1/ 1/06/C&HE/	0.40415	1.590	-0.020	-0.452
1/ 1/ 1/ 1/ 1/ 1/ 1/07/C&HE/	0.46626	1.565	-0.008	-0.357
1/ 1/ 1/ 1/ 1/ 1/ 1/08/C&HE/	0.40267	1.519	-0.015	-0.505
1/ 1/ 1/ 1/ 1/ 1/ 1/09/C&HE/	0.22597	1.234	-0.067	-0.688
1/ 1/ 1/ 1/ 1/ 1/ 1/10/C&HE/	0.08211	0.878	-0.185	-1.021
1/ 1/ 1/ 1/ 1/ 1/ 1/11/C&HE/	0.01404	0.468	-0.479	-1.717
1/ 1/ 1/ 1/ 1/ 1/ 1/12/C&HE/	0.00004	0.024	-0.841	-2.169
1/ 1/ 1/ 1/ 1/ 1/ 1/13/TOTE/	1.885		-4.230	
1/ 1/ 1/ 1/ 1/ 1/ 1/14/MAXE/		1.590		-2.540

Heat Gain and Heat Loss

Row													
1/ 1/ 1/ 1/ 1/ 1/01/HEAT/	-0.542	0.000	0.000	0.000	-0.063	-0.588	0.080	0.000	0.208	0.000	0.000	-0.904	0.000
1/ 1/ 1/ 1/ 1/ 1/01/SENC/	0.000	0.000	0.000	0.000	0.000	0.000	0.000	0.000	0.000	0.000	0.000	0.000	0.000
1/ 1/ 1/ 1/ 1/ 1/01/LATC/				0.000				0.000		0.000	0.000		0.000
1/ 1/ 1/ 1/ 1/ 1/02/HEAT/	-0.474	0.000	0.000	0.000	-0.052	-0.517	0.103	0.000	0.175	0.000	0.000	-0.766	0.000
1/ 1/ 1/ 1/ 1/ 1/02/SENC/	-0.002	0.000	0.000	0.000	-0.002	-0.002	0.002	0.000	0.003	0.000	0.000	0.000	0.000
1/ 1/ 1/ 1/ 1/ 1/02/LATC/				0.000				0.000		0.000	0.000		0.000
1/ 1/ 1/ 1/ 1/ 1/03/HEAT/	-0.417	0.000	0.000	0.000	-0.024	-0.469	0.146	0.000	0.185	0.000	0.000	-0.578	0.000
1/ 1/ 1/ 1/ 1/ 1/03/SENC/	-0.009	0.000	0.000	0.000	-0.001	-0.013	0.014	0.000	0.013	0.000	0.000	-0.004	0.000
1/ 1/ 1/ 1/ 1/ 1/03/LATC/				0.000				0.000		0.000	0.000		0.000
1/ 1/ 1/ 1/ 1/ 1/04/HEAT/	-0.216	0.000	0.000	0.000	-0.010	-0.243	0.106	0.000	0.100	0.000	0.000	-0.264	0.000
1/ 1/ 1/ 1/ 1/ 1/04/SENC/	-0.037	0.000	0.000	0.000	-0.003	-0.062	0.093	0.000	0.077	0.000	0.000	0.068	0.000
1/ 1/ 1/ 1/ 1/ 1/04/LATC/				0.000				0.000		0.000	0.000		0.000
1/ 1/ 1/ 1/ 1/ 1/05/HEAT/	-0.099	0.000	0.000	0.000		-0.111	0.063	0.000	0.044	0.000	0.000	-0.103	0.000
1/ 1/ 1/ 1/ 1/ 1/05/SENC/	-0.026	0.000	0.000	0.000		-0.071	0.184	0.000	0.130	0.000	0.000	0.217	0.000
1/ 1/ 1/ 1/ 1/ 1/05/LATC/				0.000				0.000		0.000	0.000		0.000
1/ 1/ 1/ 1/ 1/ 1/06/HEAT/	-0.024	0.000	0.000	0.000		-0.026	0.018	0.000	0.012	0.000	0.000	-0.020	0.000
1/ 1/ 1/ 1/ 1/ 1/06/SENC/	0.029	0.000	0.000	0.000		-0.028	0.246	0.000	0.156	0.000	0.000	0.404	0.000
1/ 1/ 1/ 1/ 1/ 1/06/LATC/				0.000				0.000		0.000	0.000		0.000
1/ 1/ 1/ 1/ 1/ 1/07/HEAT/	-0.011	0.000	0.000	0.000		-0.012	0.011	0.000	0.005	0.000	0.000	-0.008	0.000
1/ 1/ 1/ 1/ 1/ 1/07/SENC/	0.053	0.000	0.000	0.000		-0.008	0.263	0.000	0.159	0.000	0.000	0.466	0.000
1/ 1/ 1/ 1/ 1/ 1/07/LATC/				0.000				0.000		0.000	0.000		0.000
1/ 1/ 1/ 1/ 1/ 1/08/HEAT/	-0.019	0.000	0.000	0.000		-0.020	0.013	0.000	0.011	0.000	0.000	-0.015	0.000
1/ 1/ 1/ 1/ 1/ 1/08/SENC/	0.035	0.000	0.000	0.000		-0.014	0.208	0.000	0.174	0.000	0.000	0.403	0.000
1/ 1/ 1/ 1/ 1/ 1/08/LATC/				0.000				0.000		0.000	0.000		0.000

Row												
1/ 1/ 1/ 1/ 1/ 1/09/HEAT/	-0.063	0.000	0.000	0.000	0.000	0.000	0.032	0.000	0.032	0.000	0.000	-0.067
1/ 1/ 1/ 1/ 1/ 1/09/SENC/	-0.010	0.000	0.000	0.000	-0.044	0.142	0.000	0.138	0.000	0.226		
1/ 1/ 1/ 1/ 1/ 1/09/LATC/	0.000	0.000	0.000	0.000	0.000	0.000	0.000	0.000	0.000			
1/ 1/ 1/ 1/ 1/ 1/10/HEAT/	-0.156	0.000	0.000	-0.068	-0.165	0.052	0.000	0.084	0.000	-0.185		
1/ 1/ 1/ 1/ 1/ 1/10/SENC/	-0.044	0.000	0.000	-0.065	0.078	0.000	0.113	0.000	0.082			
1/ 1/ 1/ 1/ 1/ 1/10/LATC/	0.000	0.000	0.000	0.000	0.000	0.000	0.000					
1/ 1/ 1/ 1/ 1/ 1/11/HEAT/	-0.331	0.000	-0.017	-0.357	0.070	0.000	0.155	0.000	-0.479			
1/ 1/ 1/ 1/ 1/ 1/11/SENC/	-0.016	0.000	-0.001	-0.021	0.016	0.000	0.036	0.000	0.014			
1/ 1/ 1/ 1/ 1/ 1/11/LATC/	0.000	0.000	0.000	0.000	0.000	0.000						
1/ 1/ 1/ 1/ 1/ 1/12/HEAT/	-0.511	0.000	-0.052	-0.542	0.067	0.000	0.197	0.000	-0.841			
1/ 1/ 1/ 1/ 1/ 1/12/SENC/	-0.001	0.000	0.000	-0.001	0.000	0.000	0.001	0.000	0.000			
1/ 1/ 1/ 1/ 1/ 1/12/LATC/	0.000	0.000	0.000	0.000	0.000							
1/ 1/ 1/ 1/ 1/ 1/13/THEA/	-2.863	0.000	-0.218	-3.117	0.760	0.000	1.208	0.000	-4.230			
1/ 1/ 1/ 1/ 1/ 1/13/TSEN/	-0.027	0.000	-0.005	-0.328	1.245	0.000	1.000	0.000	1.885			
1/ 1/ 1/ 1/ 1/ 1/13/TLAT/	0.000	0.000	0.000	0.000	0.000							

Daylight

1/	1/	1/	1/	1/01/DLGT/	5.6	7.0	4.4	0.0
1/	1/	1/	1/	1/02/DLGT/	8.2	10.3	5.7	0.0
1/	1/	1/	1/	1/03/DLGT/	11.9	14.8	8.0	0.0
1/	1/	1/	1/	1/04/DLGT/	16.3	19.7	10.7	0.0
1/	1/	1/	1/	1/05/DLGT/	21.3	25.3	12.3	0.0
1/	1/	1/	1/	1/06/DLGT/	21.6	25.2	13.4	0.0
1/	1/	1/	1/	1/07/DLGT/	22.8	26.5	13.7	0.0
1/	1/	1/	1/	1/08/DLGT/	19.3	23.1	12.1	0.0
1/	1/	1/	1/	1/09/DLGT/	14.6	18.0	9.7	0.0
1/	1/	1/	1/	1/10/DLGT/	10.6	13.2	7.0	0.0
1/	1/	1/	1/	1/11/DLGT/	6.3	7.9	4.6	0.0
1/	1/	1/	1/	1/12/DLGT/	4.2	5.3	3.5	0.0
1/	1/	1/	1/	1/13/DLGT/	13.7	16.5	9.1	0.0

Heating and Cooling

2/ 1/ 1/ 1/ 1/ 1/01/C&HE/	0.54918	1.667	0.000	-0.097
2/ 1/ 1/ 1/ 1/ 1/02/C&HE/	0.52573	1.838	-0.001	-0.196
2/ 1/ 1/ 1/ 1/ 1/03/C&HE/	0.70088	1.974	0.000	0.000
2/ 1/ 1/ 1/ 1/ 1/04/C&HE/	0.74285	2.122	0.000	0.000
2/ 1/ 1/ 1/ 1/ 1/05/C&HE/	0.82044	2.212	0.000	0.000
2/ 1/ 1/ 1/ 1/ 1/06/C&HE/	0.77609	2.187	0.000	0.000
2/ 1/ 1/ 1/ 1/ 1/07/C&HE/	0.78789	2.103	0.000	0.000
2/ 1/ 1/ 1/ 1/ 1/08/C&HE/	0.74797	2.164	0.000	0.000
2/ 1/ 1/ 1/ 1/ 1/09/C&HE/	0.63591	1.844	0.000	0.000
2/ 1/ 1/ 1/ 1/ 1/10/C&HE/	0.62744	1.776	0.000	0.000
2/ 1/ 1/ 1/ 1/ 1/11/C&HE/	0.48027	1.585	-0.001	-0.103
2/ 1/ 1/ 1/ 1/ 1/12/C&HE/	0.51359	1.567	-0.001	-0.158
2/ 1/ 1/ 1/ 1/ 1/13/TOTE/	7.908		-0.003	
2/ 1/ 1/ 1/ 1/ 1/14/MAXE/		2.212		-0.196

Heat Gain and Heat Loss

2/ 1/ 1/ 1/ 1/01/HEAT/	-0.001	0.000	0.000	0.000		-0.001	0.000	0.000	0.000
2/ 1/ 1/ 1/ 1/01/SENC/	-0.107	0.000	0.000	0.000	0.005	0.053	0.202	0.182	0.549
2/ 1/ 1/ 1/ 1/01/LATC/				0.027					0.027
2/ 1/ 1/ 1/ 1/02/HEAT/	-0.002	0.000	0.000	0.000		-0.002	0.002	0.002	-0.001
2/ 1/ 1/ 1/ 1/02/SENC/	0.108	0.000	0.000	0.000	0.004	0.050	0.211	0.152	0.526
2/ 1/ 1/ 1/ 1/02/LATC/				0.019					0.019
2/ 1/ 1/ 1/ 1/03/HEAT/	0.000	0.000	0.000	0.000		0.000	0.000	0.000	0.000
2/ 1/ 1/ 1/ 1/03/SENC/	0.162	0.000	0.000	0.000	0.003	0.092	0.272	0.172	0.701
2/ 1/ 1/ 1/ 1/03/LATC/				0.013					0.013
2/ 1/ 1/ 1/ 1/04/HEAT/	0.000	0.000	0.000	0.000		0.000	0.000	0.000	0.000
2/ 1/ 1/ 1/ 1/04/SENC/	0.174	0.000	0.000	0.000	0.004	0.095	0.313	0.157	0.743
2/ 1/ 1/ 1/ 1/04/LATC/				0.019					0.019
2/ 1/ 1/ 1/ 1/05/HEAT/	0.000	0.000	0.000	0.000		0.000	0.000	0.000	0.000
2/ 1/ 1/ 1/ 1/05/SENC/	0.188	0.000	0.000	0.000		0.101	0.371	0.161	0.820
2/ 1/ 1/ 1/ 1/05/LATC/									
2/ 1/ 1/ 1/ 1/06/HEAT/	0.000	0.000	0.000	0.000		0.000	0.000	0.000	0.000
2/ 1/ 1/ 1/ 1/06/SENC/	0.160	0.000	0.000	0.000		0.081	0.375	0.160	0.776
2/ 1/ 1/ 1/ 1/06/LATC/	0.000	0.000	0.000	0.000		0.000	0.000	0.000	0.000
2/ 1/ 1/ 1/ 1/07/HEAT/	0.000	0.000	0.000	0.000		0.000	0.000	0.000	0.000
2/ 1/ 1/ 1/ 1/07/SENC/	0.165	0.000	0.000	0.000		0.085	0.381	0.156	0.788
2/ 1/ 1/ 1/ 1/07/LATC/	0.000	0.000	0.000	0.000		0.000	0.000	0.000	0.000
2/ 1/ 1/ 1/ 1/08/HEAT/	0.000	0.000	0.000	0.000		0.000	0.000	0.000	0.000
2/ 1/ 1/ 1/ 1/08/SENC/	0.162	0.000	0.000	0.000		0.083	0.333	0.170	0.748
2/ 1/ 1/ 1/ 1/08/LATC/	0.000	0.000	0.000	0.000		0.000	0.000	0.000	0.000

Row									
2/ 1/ 1/ 1/ 1/ 1/09/HEAT/	0.000	0.000	0.000	0.000	0.000	0.000	0.000	0.000	0.000
2/ 1/ 1/ 1/ 1/ 1/09/SENC/	0.136	0.000	0.000	0.064	0.285	0.000	0.151	0.000	0.636
2/ 1/ 1/ 1/ 1/ 1/09/LATC/	0.000	0.000	0.000	0.000	0.000	0.000	0.000	0.000	0.000
2/ 1/ 1/ 1/ 1/ 1/10/HEAT/	0.000	0.000	0.000	0.000	0.000	0.000	0.000	0.000	0.000
2/ 1/ 1/ 1/ 1/ 1/10/SENC/	0.136	0.000	0.000	0.063	0.257	0.000	0.172	0.000	0.627
2/ 1/ 1/ 1/ 1/ 1/10/LATC/	-0.001	0.000	0.000	-0.001	-0.001	0.000	0.000	0.000	-0.001
2/ 1/ 1/ 1/ 1/ 1/11/HEAT/	-0.001	0.000	0.000	-0.001	-0.001	0.000	0.000	0.000	-0.001
2/ 1/ 1/ 1/ 1/ 1/11/SENC/	0.088	0.000	0.001	0.028	0.196	0.000	0.167	0.000	0.480
2/ 1/ 1/ 1/ 1/ 1/11/LATC/	-0.001	0.000	0.007			0.000		0.000	0.007
2/ 1/ 1/ 1/ 1/ 1/12/HEAT/	0.103	0.000	0.000	-0.001	-0.001	0.000	0.000	0.000	-0.001
2/ 1/ 1/ 1/ 1/ 1/12/SENC/	-0.001	0.000	0.004	0.046	0.187	0.000	0.174	0.000	0.514
2/ 1/ 1/ 1/ 1/ 1/12/LATC/	0.088	0.000	0.017			0.000		0.000	0.017
2/ 1/ 1/ 1/ 1/ 1/13/THEA/	-0.006	0.000	0.000	-0.006	-0.004	0.000	0.004	0.000	-0.003
2/ 1/ 1/ 1/ 1/ 1/13/TSEN/	1.688	0.000	0.022	0.841	3.383	0.000	1.974	0.000	7.908
2/ 1/ 1/ 1/ 1/ 1/13/TLAT/		0.000	0.101			0.000		0.000	0.101

Day	Light			
2/ 1/ 1/ 1/ 1/ 1/ 1/01/DLGT/	17.1	21.4	11.2	0.0
2/ 1/ 1/ 1/ 1/ 1/ 1/02/DLGT/	20.7	25.9	13.6	0.0
2/ 1/ 1/ 1/ 1/ 1/ 1/03/DLGT/	23.6	29.4	15.5	0.0
2/ 1/ 1/ 1/ 1/ 1/ 1/04/DLGT/	25.6	31.3	17.8	0.0
2/ 1/ 1/ 1/ 1/ 1/ 1/05/DLGT/	27.1	33.1	17.3	0.0
2/ 1/ 1/ 1/ 1/ 1/ 1/06/DLGT/	25.7	31.1	16.3	0.0
2/ 1/ 1/ 1/ 1/ 1/ 1/07/DLGT/	26.3	31.9	17.6	0.0
2/ 1/ 1/ 1/ 1/ 1/ 1/08/DLGT/	25.9	31.6	17.9	0.0
2/ 1/ 1/ 1/ 1/ 1/ 1/09/DLGT/	24.1	29.7	16.9	0.0
2/ 1/ 1/ 1/ 1/ 1/ 1/10/DLGT/	21.7	27.0	15.3	0.0
2/ 1/ 1/ 1/ 1/ 1/ 1/11/DLGT/	17.5	21.8	11.5	0.0
2/ 1/ 1/ 1/ 1/ 1/ 1/12/DLGT/	15.6	19.5	10.3	0.0
2/ 1/ 1/ 1/ 1/ 1/ 1/13/DLGT/	22.6	27.9	15.0	0.0

APPENDIX F

EXAMPLE OF ACCESS DATABASE

File Edit View Insert Format Records Tools Window Help

Field1	Field2	Field3	Field4	Field5	Field6	Field7	Field8	Field9
1/ 1/ 1/ 1/ 1/ 1/	0.00133	0.137	-0.785	-2.499	128	0.405		
1/ 1/ 1/ 1/ 1/ 1/	5.5	6.9	4.4	0				
1/ 1/ 1/ 1/ 1/ 1/	-0.533	0	0	0	-0.077	-0.58	0.077	0
1/ 1/ 1/ 1/ 1/ 1/					0			0
1/ 1/ 1/ 1/ 1/ 1/	-0.009	0	0	0	-0.001	-0.01	0.003	0
1/ 1/ 1/ 1/ 1/ 1/	0.00356	0.438	-0.666	-2.378	109	0.405		
1/ 1/ 1/ 1/ 1/ 1/	8.1	10.2	5.7	0				
1/ 1/ 1/ 1/ 1/ 1/	-0.463	0	0	0	-0.064	-0.506	0.096	0
1/ 1/ 1/ 1/ 1/ 1/					0			0
1/ 1/ 1/ 1/ 1/ 1/	-0.014	0	0	0	-0.002	-0.016	0.009	0
1/ 1/ 1/ 1/ 1/ 1/	0.02382	0.628	-0.474	-1.971	121	0.405		
1/ 1/ 1/ 1/ 1/ 1/	11.7	14.6	7.9	0				
1/ 1/ 1/ 1/ 1/ 1/	-0.387	0	0	0	-0.028	-0.433	0.12	0
1/ 1/ 1/ 1/ 1/ 1/					0			0
1/ 1/ 1/ 1/ 1/ 1/	-0.04	0	0	0	-0.003	-0.051	0.039	0
1/ 1/ 1/ 1/ 1/ 1/	0.12791	1.256	-0.211	-1.156	109	0.399		
1/ 1/ 1/ 1/ 1/ 1/	16.1	19.4	10.5	0				
1/ 1/ 1/ 1/ 1/ 1/	-0.186	0	0	0	-0.011	-0.207	0.08	0
1/ 1/ 1/ 1/ 1/ 1/					0			0
1/ 1/ 1/ 1/ 1/ 1/	-0.067	0	0	0	-0.006	-0.1	0.119	0
1/ 1/ 1/ 1/ 1/ 1/	0.30877	1.755	-0.079	-0.841	107	0.395		
1/ 1/ 1/ 1/ 1/ 1/	21	25	12.1	0				
1/ 1/ 1/ 1/ 1/ 1/	-0.084	0	0	0	0	-0.092	0.047	0
1/ 1/ 1/ 1/ 1/ 1/					0			0
1/ 1/ 1/ 1/ 1/ 1/	-0.042	0	0	0	0	-0.091	0.2	0
1/ 1/ 1/ 1/ 1/ 1/	0.51027	1.888	-0.013	-0.379	103	0.394		
1/ 1/ 1/ 1/ 1/ 1/	21.3	24.8	13.2	0				
1/ 1/ 1/ 1/ 1/ 1/	-0.019	0	0	0	0	-0.02	0.013	0

Record: 14 ◄ 1 ► ►I ►* of 675840

Datasheet View   NUM

BIBLIOGRAPHY

Akin, O., "How do architects design?," *Artificial Intelligence and Pattern Recognition in Computer Aided Design*, Latombe, J.C. (ed), North Holland, 1978.

Baker, N., and Steemers, K., "Energy and Environment in Architecture", A Technical Design Guide, E&FN Spon, London, 2000.

Bellchambers, H.E., Lambert, G.K. and Ruff, H.R., "Modern Aids to Lighting Design - Computer Techniques", *Trans Illumination Engineer Society*, Vol. 26, No. 3, pp.107-122, 1961.

Boonyathikarn, S., "Technique for Designing Energy Efficient House", Chulalongkong University Press, Bangkok, 1999.

Brittain, J., "The Cost of Oversized Plant", BSRIA Guidance Note, No. 12, *Building Energy Analysis*, Simulation Research Group, Lawrence Berkley National Laboratory(LBNL), 1997.

Chou, S.K. and W.L. C., "Development of an Energy Estimating Equation for Large Commercial Buildings", *International Journal of Energy Research*, Vol. 17, pp. 759-773, 1993.

Chou, S.K. and Y.W. W., "Prediction Energy Performance of Commercial Buildings in Singapore," *ASHRAE Transactions*, Vol. 92. Part.1A, pp. 116-136, 1986.

CIBSE, "CIBSE Guide, Volume A - Design Data", *The Chartered Institute of Building Services Engineers*, Staples Printers, St Albans, England, 1986.

Condit, Carl W., "Chicago 1930-70, Building, Planning, and Urban Technology", The University of Chicago Press, 1974.

Cowan H.J., "An Historical Outline of Architectural Science", Elsevier, Amsterdam, 1966.

Curtis, R.B., Birdsall, B., Buhl, W.F., Erdem, E., Eto, J., Hirsch, J.J., Olson, K., and Winkelmann, F.C., "The use of DOE-2 to Evaluate the Energy Performance of Buildings," *Proceedings of ASEAN Conference on Energy Conservation in Buildings*, Development and Building Control Division, PWD, Singapore, 1984.

Hart, S., "Architecture", V.87, No.8, pp.116 (4), August 1998.

Howard, H., "Pipe Sizing with an Electronic Computer", *JIHVE,* Vol..28, No.1, pp..23, 1960.

238

Kusuda, T. and Sud I., "Updated ASHRAE TC 4.7 Simplified Energy Analysis Procedure", *ASHRAE Journal*, July 1982.

Lam, Joseph C., and C.M.Hui S., "Sensitivity Analysis of Energy Performance of Office Buildings", *Building and Environment*, Vol. 31, No.1, pp..27-39, 1996.

Lam, Joseph C., C.M. Hui S., and L.S.Chan A., "Regression analysis of high-rise fully air-conditioned office buildings", *Energy and Buildings*, Vol. 26, pp.189-197, 1996.

Longmore, J., "BRS Daylight Protractors", Department of Environment Building Research Station, Her Majesty's Stationary Office (HMSO), London, 1968.

Manning P., "Environmental Evaluation", *Building and Environment*, Vol.22 No. 3, pp.201-208, 1987.

Olgyay, V., "Design with Climate: Bio-climatic Approach to Architectural Regionalism", Princeton University Press, 1963.

Parsloe, C.J., "Over-engineering in Building Services: An International Comparison of Design and Installation Methods", *BSRIA Technical Report*, No. TR21/95, BSRIA, 1995.

Race, G.L., "Design Margins in Building Services", *Building Services Journal*, August 1997.

Ruyssevelt, P., Batholemew, D., "Stimulating Simulation, Computer Tools in Building Services", *Building Services Journal*, October 1997.

Sullivan, R.T. and Nozaki S.A., "Multiple Regression Techniques applied to Fenestration Effects on Commercial Building Energy Performance", *ASHRAE Transactions*, Vol. 90, 1984.

Sullivan, R.T., Nozaki S.A., Johnson R., and Selkowitz S., "Commercial Building Energy Performance Analysis using Multiple Regression", *ASHRAE Transactions*, Vol. 91, 1985.

Szokolay, S.V., "Environmental Science Handbook", The Construction Press Ltd, Lancaster, England, 1980.

Szokolay, S.V., "Thermal Design of Buildings", RAIA Education Division, Canberra, Australia, 1987.

Szokolay, S.V., "Solar Geometry", *PLEA Note No. 1*, University of Queensland Department of Architecture, 1996.

Tufte, E.R., "The Visual Display of Quantitative Information", The Graphic Press: Cheshire, Connecticut, 1984.

Tufte E.R., "Visual Explanation, Images and Quantities, Evidence and Narrative", The Graphic Press: Cheshire, Connecticut, 1995.

Tufte E.R., "Envision Information", The Graphic Press: Cheshire, Connecticut, 1995.

Turiel, I., Richard B., Mark S., and Mark L., "Simplified Energy Analysis Methodology for Commercial Building", *Energy and Building*, Vol. 6, pp.67-83, 1984.

Wilcox, B.A., "Development of the Envelope Load Equation for ASHRAE Standard 90.1", *ASHRAE Transactions*, Vol. 97, 1991.